Homesick

CRITICAL GLOBAL HEALTH: Evidence, Efficacy, Ethnography
A series edited by Vincanne Adams and João Biehl

Duke University Press   *Durham and London*   2025

Homesick

NICHOLAS SHAPIRO

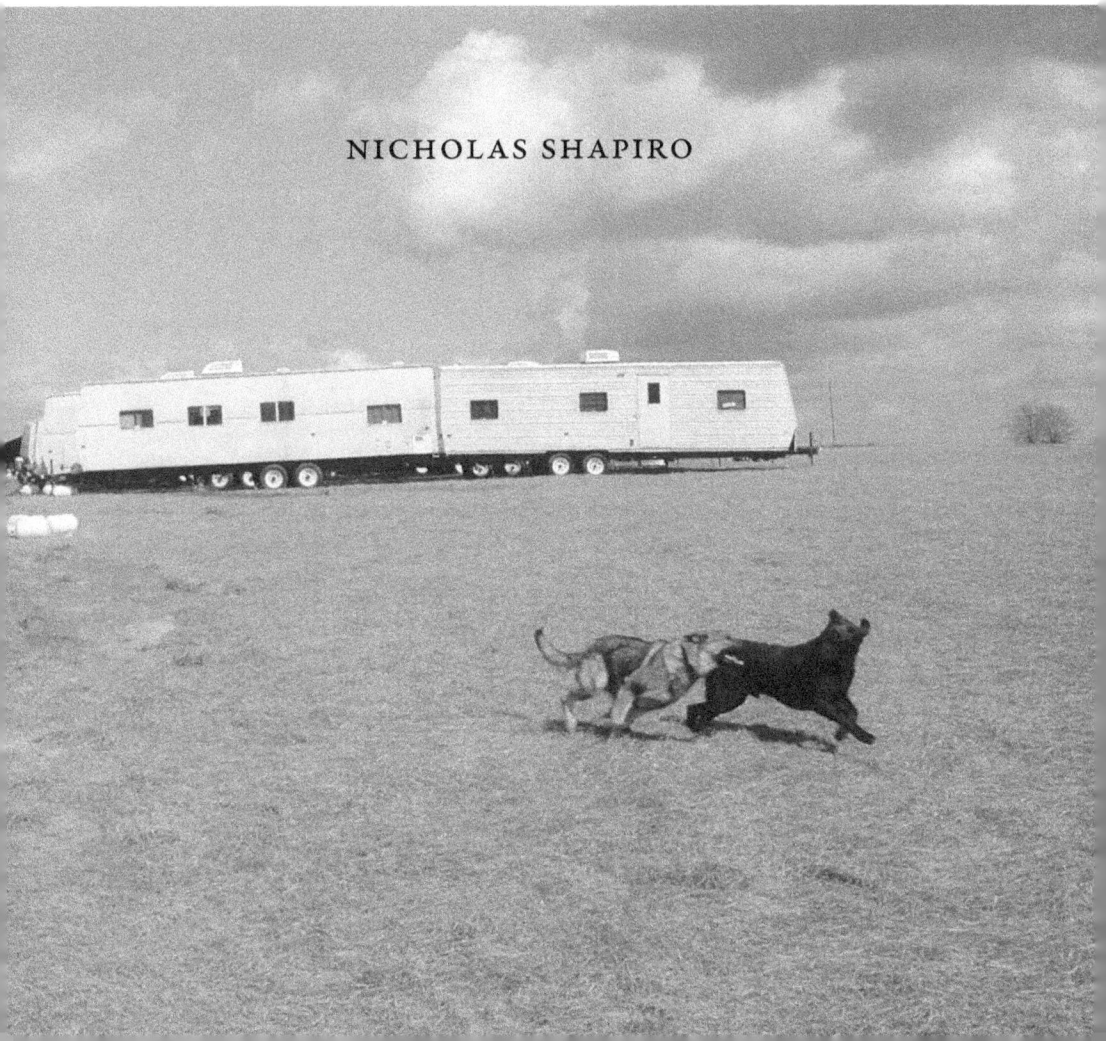

I've been having this Homesick feeling,
even though I am not sure what home I am sick for anymore. . . .

MONIQUE VERDIN, director of the Land Memory Bank
and Seed Exchange, former member of the United
Houma Nation Tribal Council, August 29, 2022

Project Editor: Bird Williams
Designed by Courtney Leigh Richardson
Typeset in Garamond Premier Pro and Wrangler
by Westchester Publishing Services

Library of Congress Cataloging-in-Publication Data
Names: Shapiro, Nicholas Edward, author.
Title: Homesick / Nicholas Edward Shapiro.
Other titles: Critical global health.
Description: Durham : Duke University Press, 2025. | Series: Critical
global health: evidence, efficacy, ethnography | Includes bibliographical
references and index.
Identifiers: LCCN 2024060619 (print)
LCCN 2024060620 (ebook)
ISBN 9781478032441 (paperback)
ISBN 9781478029076 (hardcover)
ISBN 9781478061298 (ebook)
ISBN 9781478094357 (ebook other)
Subjects: LCSH: United States. Federal Emergency Management
Agency. | Emergency housing—Health aspects—Gulf Coast (U.S.) |
Mobile homes—Health aspects—Gulf Coast (U.S.) | Mobile home
industry—Health aspects—Gulf Coast (U.S.) | Formaldehyde—
Toxicology—Gulf Coast (U.S.) | Housing and health—Gulf
Coast (U.S.) | Biopolitics—United States. | United States—Social
conditions—21st century.
Classification: LCC HV555. U62 G2025 (print)
LCC HV555. U62 (ebook)
DDC 613/.50976—dc23/eng/20250520
LC record available at https://lccn.loc.gov/2024060619
LC ebook record available at https://lccn.loc.gov/2024060620

Cover art: copyright Akasha Rabut, https://www.akasharabut.com.

This book is freely available in an open access edition thanks to TOME (Toward
an Open Monograph Ecosystem)—a collaboration of the Association of American
Universities, the Association of University Presses, and the Association of Research
Libraries—and the generous support of Arcadia, a charitable fund of Lisbet Rausing
and Peter Baldwin, and the UCLA Library. Learn more at the TOME website, available
at openmonographs.org.

publication supported by a grant from
The Community Foundation for Greater New Haven
as part of the URBAN HAVEN PROJECT

CONTENTS

PREFACE

Our No. 1 priority is housing, our No. 2 priority is housing,
and after that, at No. 3, we'd put housing.

GULF COAST DIRECTOR FOR FEMA

Formaldehyde is the only thing keeping me alive!

NEW ORLEANS BLUES SINGER
BENNY MAYGARDEN

We sat in folding chairs facing the street. A thick chain dangled from the live oak branch above us. My companion, Jeremy, looked up into the tangle of thick limbs and reminisced about when he had the energy to tinker with cars—using the metal rigging above to pull out engines. As he lowered his gaze from the tree's canopy back down to his trailer, jacked up on cinder blocks to our left, his stories of grease and metal ground to a halt. It had been years since he felt well. He, his wife, and their daughter, all Black, have been living in a twenty-one-foot-long travel trailer in the Mid-City neighborhood of New Orleans for the past five years.[1] Their trailer and 120,000 nearly identical trailers had been issued by the Federal Emergency Management Agency (FEMA) to those residents left without a habitable home after Hurricanes Katrina and Rita in 2005. Clearing phlegm from his throat, he recollected in his slightly high-pitched voice: "When we moved in, it was kinda, it was kinda . . . you could really *feel* the formaldehyde. We had to open the windows or we had to run the air-conditioning. . . . The formaldehyde would burn your eyes. It would burn your nostrils." Although the chemical vapors became less intense or at least less noticeable over time, his family's upper airways were constantly inflamed. Bronchitis, chronic fatigue, nasal congestion, insomnia, and eye irritation all became features of their constitution. He laughed when I asked him how many over-the-counter medications he used in the hopes of feeling like he did before he moved in. "Aw, man. . . . Lemme see: Robitussin PM, sinus tablets, Theraflu™ Severe Cold and Cough, NyQuil™, Alaway™, Wal-itin . . . probably about seven or eight different types of medications. Some for the cough, some for the sinus and allergy, some for the, um, trying to get the mucus up, ya know?" Inside his trailer, the windowsills and door frames were lined with the small cardstock boxes of over-the-counter medications. Plastic pillboxes were stuffed between the cushions of the standard couch found in all such trailers. An asthma inhaler, which he said was prescribed last week, lay on top of the TV, and an empty blister pack of decongestant tablets sat atop discarded coupons in the trash can.

After the destruction of their rented suburban home, the family moved to the city and placed their trailer on a church friend's vacant lot. Friends and family scattered across the region, and new neighbors were all living in FEMA trailers. They struck up friendships with their neighbors, many of whom also soon began to suffer from similar maladies. "We communicated a lot," he recounted. "When they went to the doctor, they would tell me what's going on with 'em and what medicines they picked up." Together, old friends and new neighbors tried to sleuth out the source of their shared symptoms. With resources tight and their new mass-produced homes identical, information from an individual's doctor visit might reveal something that would help everyone. Their uniform homes patterned their days and aligned their maladies, bringing their bodies into a near (dysfunctional) synchronicity. From the engineered wood that gave form to most surfaces in these trailers, toxic chemicals, notably formaldehyde, were continuously wriggling free of their bonds and into breathing space in a process known as "off-gassing." This temperature-dependent process would increase as the sun rose in the morning, forcing inhabitants outside during the high off-gassing periods of the early afternoon. These converging homesicknesses—of chemical and domestic dispossession, of the sick markets that reap profits from racially patterned exposures, and of the struggle to move toward a home that supports life rather than draining it—all lie at the center of this book.

In time, the book will move well beyond the Gulf Coast and Katrina. But it starts here as an inroad to more general questions of how homes in America became sick and what can be done about it. Quiet caustic chemical exposures that strike in the privacy of one's own home often cultivate profound somatic and social alienation. Restorative retreat is impossible in these toxic homes scattered across the country. It's often unclear where one should look for accountability, which other households might also be afflicted, and who could help them figure out the cause of their deteriorating body. With little training in environmental illnesses, clinicians often quickly dismiss individual cases as anything but chronic encounters with chemicals in the home. But after Katrina, when hundreds of thousands of survivors synchronously filed into nearly identical toxic homes, illnesses that manifest individually and often make people feel more isolated instead pulled into transitory alignment. The FEMA trailers and the sicknesses they spurred created a fleeting social dynamic across an entire region, garnering attention from across the world for otherwise commonplace domestic toxicity.

This minor way of being difunctionally in common was easily overlooked among the disorienting postdisaster landscape, the multiple struggles of rebuilding (cf. Adams 2013), and the big rhetoric of postdisaster community resilience (Solnit 2009: chapter 5; Flaherty 2010).[2] Yet, these emergency housing units and the toxic gas emanating from their building materials undergirded the everyday lives of a Gulf Coast in recovery, creating an anomalous moment for barely perceptible chemical exposures to be reckoned with. These toxic trailers produced international recognition of domestic chemical risks, multiple congressional hearings, and a historic, albeit regressive, shift in the way the building materials for all US homes are regulated.

This book will also be attentive to what this rare media and regulatory attention did not and perhaps cannot do. But before delving into how I spent more than a decade tracking these trailers as they were quasi-legally resold across the country, before drawing lines of connection from these boxy temporary homes to homes across the country, before detailing a cascade of concepts for reckoning with the experience of chronic exposure, articulating an array of small enfeebling encounters with chemicals into understandings of systematic harm, and putting multiple theories of change to the test, I want to set the scene more vividly.

In this preface, I document the clouds of formaldehyde exposures that emanated from the engineered woods of trailer walls, floors, doors, ceilings, and cabinetry, alongside a more general aura of horror that lingered in the city among flotsam and vacant houses. In the spatial shake-up of destruction and recovery, these mass-produced, bare-bones homes became a distributed city of their own, grafted atop a ravaged region.

IN THE EARLY HOURS of August 29, 2005, Hurricane Katrina churned across the extreme southeastern tip of Louisiana, an alluvial peninsula that could be thought of as a toe in a boot-shaped state. With wind speeds of more than 120 miles an hour and a diameter of four hundred miles, the hurricane veered east of New Orleans, making final landfall near the Louisiana-Mississippi border. Coastal Mississippi suffered near-total destruction. The storm carved its way north through Mississippi, spawning dozens of tornadoes and dissolving into heavy rain over the Midwest. Simultaneously, a surge of seawater—propelled by the hurricane as it made landfall in Mississippi—swept west, toward New Orleans. Two long concrete barriers, built to shield the city from flooding, funneled a broad swell of water into a concentrated torrent that tore through the mouth of a manufactured shipping channel and into the city's inner harbor and canals. Amplified in intensity by the very infrastructure designed to protect the city, the surge compromised New Orleans's levee system in over fifty locations. This paradox, of protection amplifying exposure, would continue to echo through the recovery and throughout this book. Eighty percent of New Orleans lay submerged. Less than a month after Katrina inundated the central Gulf Coast, Hurricane Rita careened up the Texas-Louisiana border.[3] The storm upended smaller coastal communities and pushed the overwhelmed federal response even closer toward its breaking point.

After the storms, the walls of FEMA trailers were entrusted with the gargantuan task of reinstating the spatial syntax of the Gulf Coast, of separating ruins from residence, of carving out little pockets of order from the chaos of mass destruction.[4] The disaster reordered life and landscape at a magnitude that is difficult to conceive.[5] After more than 130 billion gallons of water poured into New Orleans, the city lay almost entirely vacant for months. Mold colonies burrowed deep into the frames of houses and cat's claw, a rapidly climbing vine with yellow spring flowers, began to net the city's blighted buildings. Long before they became a symbol of a protracted and painful recovery, FEMA trailers were cornerstones of the infrastructure of hope. They were gleaming white boxes that rode into the dismal postdisaster landscape on trains and behind trucks. For the 1.36 million diasporic Gulf Coast residents who applied to FEMA for individual assistance within a month of Katrina's landfall, the trailers were beacons of the possibility of reassembling a livable life.

The State Department of Economic Development established an Early Return Program and issued trailers to those workers deemed essential to the city's recovery. Through this program, a white architect in the relatively affluent Lake Terrace neighborhood, on the northern edge of City Park by Lake Pontchartrain, was granted early access to his property. Years later, seated around the conference table at his Central Business District office, he recalled his commute home from work, through a city dotted with powerless streetlamps. Every night for months, he would slowly drive down pitch-black, rubble-lined streets with no signs of other living humans. Even as trailers began to be deployed en masse, four to six months after inundation, the tone of the city oscillated between eerie, foreign, and wild. "It feels like there are coyotes out there," a middle-aged Black mother in the Hollygrove neighborhood recollected thinking early in the recovery, when strange and frightening noises wafted through the thin walls of her trailer at night. She was unable to work, overwrought by the anxiety and depression of living in an upended city. A Central City man shut out the unthinkable state of his city when he returned home every night, after long hours of maintenance work, to the remove of a donated Carnival Cruise ship moored in the Mississippi River. After he moved into his trailer—a much larger space in the thick of destruction—he took to drinking alone, late into the night.

These diminutive emergency housing units were deployed both to the driveways of homes in ruin (106,126 households) and, for those who possessed no real estate of their own (21,105 households), to federally run trailer parks. The recovery assistance of the trailer program provided more benefit to those who owned property—specifically property that had enough space for a trailer in the front, side, or back yard. Even with homeownership, those in predominantly Black neighborhoods with tight lot spacing were often not able to receive FEMA trailers on their property due to a lack of space.

For example, after Betty (whom you will meet in chapter 4) was flooded out of her Upper Ninth Ward home, she was granted a FEMA trailer but had no place to put it on her property. She spent a couple of months bouncing between the homes of family in Lake Charles and Baton Rouge. Like many families, and particularly Black families, she was commuting back and forth from New Orleans to Baton Rouge until one of her girlfriends gave her permission to set up a FEMA trailer on her property in Boutte, Louisiana—about thirty minutes outside of New Orleans. That arrangement knocked fifty minutes off her daily commute. Betty leapt at the chance. But within a few months of moving in, Betty came down with pneumonia for the first time in her fifty-year-long life. Eight months after moving in, Betty found ballooning lumps under her breast. Since then, she has endured the constant hurdles of surgeries and chemo regimens in the battle against persistent breast cancer.

People who had previously lived in public housing would often be assigned residency in far-flung FEMA trailer parks without access to transportation, child care, medical facilities, and social services. Housing policy experts quickly deplored the ghettoizing tendency of the group-site model that "epitomized everything that housing policy can do wrong for families" (Turner 2007: 1).[6] To avoid living in trailers, middle- and upper-income owners of damaged homes

FIGURE P.I. FEMA trailer park under construction in Louisiana, late 2005. Photo by
Marvin Nauman/FEMA.

entered the already depleted New Orleans rental market, increasing the cost
of rent by 40 to 70 percent compared to pre-Katrina prices (Bernardi 2007).
Priced out of renting, lower-income households found it increasingly
difficult to locate permanent housing and languished in FEMA group sites (see
fig. P.1).

Marquisha was thirteen when she moved into a trailer cluster just north of
the Iberville housing projects that she used to call home. Her family was in the
minority of public housing residents who were able to find temporary housing
close to where they lived before the storm. Her most vivid memories from her
trailer tenure involve sprinting and diving in the door of her temporary home
to seek shelter from nearby gunfire on multiple occasions. While reassuring for
Marquisha, the thinness of walls proved deadly for others. At 4:22 a.m. on a
July morning in 2008, twenty-eight bullets passed through the aluminum siding and plywood wall of a Sheetrock contractor's FEMA trailer before striking
his sleeping body. The predictable architecture of mass-produced FEMA trailers facilitated the ease of his murder.

Marquisha's sister, her brother, a cousin, her aunt, her mother, and her father
all shared the small home with her. Mornings were particularly hectic. As the six
of them prepared for work or school, they cycled into the miniature bathroom,
through the hallway that was also the kitchen, and then crawled under the
ironing board that perennially blocked the door. Upon returning home from
school, Marquisha preferred to stay outside the trailer, as there was something
unspeakably off-putting about its indoor atmosphere: "You know how you go

into somewhere and feel like it's kinda drowsy ... your head would hurt? Every time I went up in there my head would hurt a lot." While living in the trailer, Marquisha was diagnosed with asthma and pneumonia. In that same year, her sister's eczema intensified—increasing in terms of both surface area and severity.

Long-standing social inequalities sculpted correlations between income and elevation above sea level in New Orleans, and they exposed those at the lower end of both spectrums to greater risks of flood-related displacement. As disasters often do, the flooding of the city tended to "magnify the social situations that existed before" (Miller and Rivera 2010: 180). By contrast, the impacted neighborhoods of Mississippi, which suffered a direct hit from the hurricane's winds and storm surge, were predominantly white. Wealthy neighborhoods on Mississippi's shoreline were as likely or, in a few instances, more likely to have sustained heavy damage than lower-income neighborhoods (Craemer 2010: 369). Using FEMA trailer-park populations as a proxy for all FEMA trailer residents, most FEMA trailer inhabitants in Louisiana were Black and in Mississippi most were white (Larrance et al. 2007). Across both states at the epicenter of hurricane recovery, lower-income residents disproportionately inhabited FEMA trailers and averaged a longer occupancy than their wealthy counterparts.

THE DAMAGE FROM Katrina and Rita spanned hundreds of miles of coastline. From Cambodian refugee fisherfolk in southwest Alabama to ranchers in east Texas, the white aluminum siding and gaunt interior of the trailers functioned as the common foundation on which the Gulf was rebuilt. These temporary housing units represent a truly unique moment in US housing history, as never before had so many people been thrust so abruptly into nearly identical living conditions.[7]

The FEMA trailers can be viewed as a neighborhood or city of their own—one that was distributed across the Gulf Coast, pivoting around New Orleans. An auxiliary city, it consisted of homogeneous architecture and more than 300,000 residents grafted atop a ravaged coastline. This scattered city was made in the hopes that it would be quickly disassembled. Such an infrastructural archipelago is perhaps best described as livable scaffolding, designed to assist the Gulf Coast in rearticulating discrete senses of place and community shaken apart by disaster. The entire project was stopgap medicine for a city with profound and thickening homesickness.

In unison, the cramped space of the trailers ignited arguments. Lovers broke up. Siblings fought. Thin walls allowed for the broadcasting of domestic conflicts to neighbors. At the group sites, audio awareness of nearby disputes or domestic violence made stress feel almost contagious. Others, like Lizzie (whom you will meet in chapter 2), reveled in their first taste of privacy after months spent shifting between shelters and the homes of friends and relatives. Some found the compactness of the FEMA trailer to be emotionally reassuring after experiencing total loss—there was no empty domestic space to remind them of destroyed possessions. "It was all the space I could handle," recounted one white woman who placed her trailer on her mother's property on the West Bank after her Uptown apartment was rendered uninhabitable. A Cajun man who lived in one of the last towns at the mouth of the Mississippi recounted with an ambiguously sarcastic, mischievous, and sincerely upbeat grin, "It's like living in a Swiss Army knife. What more could you ask for?" Almost everybody boasted of their ability to throw parties in their trailer, even if everyone usually ended up outside with "something of a headache." At least one trailer served as a punk music venue. Life in its full dynamism went on in the trailers, even if it was predominated by dark notes for many.

As Gulf Coast residents began to settle into their emergency housing units, the trailers became a focal point. "We were all fascinated with each other's trailer experiences. It was a big topic of conversation," noted Miranda, a white resident of the Holy Cross neighborhood of the Lower Ninth Ward. The mundane reality of residents' domestic space fueled much of their discussion. Miranda said, "It would just be stuff like 'Where did you get your propane refilled?' or 'How does your heat work?' or we would be jealous that somebody had the bump-out with extra room. And then definitely ones that would be more toxic, we would comment on them."

While life in the trailers involved navigating new common spaces and domestic technologies, some aspects of their deployment were all too familiar. Out in the farthest corner of New Orleans East—a large suburban section of the city—is a neighborhood called Versailles. Residents of Versailles, a bastion for Vietnamese Catholics fleeing the war in the 1970s, assert that they are the most concentrated Vietnamese population outside of Vietnam (Chiang 2010). In late summer 2010, I met with Ray, who worked at a community development corporation associated with Mary Queen of Vietnam Church. Fisherfolk, unemployed by the BP oil-drilling disaster, milled about outside. In a back office, he began to tell me about life in his FEMA trailer park, which had been located about a half a mile away on a vacant lot across from the church and not far from where he had lived before the storm.

His was a relatively small displacement compared to the numerous FEMA trailer parks located an hour's drive from inhabitants' prestorm residences.

Sipping from his fluorescent red slushy between sentences, Ray nonchalantly described how racial patterns of segregation between Vietnamese, Black, and white blocks were reproduced in the miniature scale of the trailer park: "It was like the neighborhood just got smaller." Elaborate community and personal gardens of several Vietnamese American residents were destroyed by the storm along with their homes. In the park, they quickly began cultivating the driveways allotted to them beside their trailers to grow Vietnamese vegetables and herbs that could not be found in local stores.

The fences and security of these trailer parks was often lamented by outside reporters trying to access the parks for stories (cf. Goodman 2006), but Ray was quite happy about the presence of security personnel and the mandatory sign-in sheets. He noted that the homogeneity of the trailers extended beyond their architecture to their ostensibly unique keys: "The keys that they made for the trailer, some of them were identical to other trailers, so you can break into other people's trailers very easily." Others also noted the key similarity problem and the common occurrence of propane theft that was reduced by the presence of security. Some 118,000 travel trailers built to FEMA specifications utilized three different types of lock; each type of lock had fifty-one, one hundred, or two hundred unique keys. In a FEMA trailer park of two hundred identical trailers, a single resident would likely be able to open at least one, and as many as four, other homes in their park without needing to force entry (Engber 2006). In the largest park of 1,600 trailers, upward of thirty-two households would share the same key.

Although group sites accounted for the vast minority of FEMA trailer placements (five times fewer than on private property), they were the almost-exclusive focus of post-Katrina media and social scientific research on these emergency housing units. The trailer density of the group sites lent themselves to door-knocking reporters and surveying academics, which made them an easier object of scrutiny than the larger constellation of FEMA trailers scattered across the region. These trailer parks also served to aggregate and, due to their often-remote locations, amplify the woes of displacement.

These trailer parks became increasingly bleak as time rolled on. By 2007, one in two Mississippi group-site residents did not possess health insurance, compared to the already high state average of 17 percent. Residents also carried a disproportionately high burden of chronic disease (Shehab et al. 2008). A 2006 survey of 229 individuals across Louisiana and Mississippi group sites found "shelter/privacy" to be the largest self-reported problem of FEMA trailer-park life (Larrance et al. 2007: 593). Shelter and privacy were the primary provisions of these temporary homes but also the primary shortcoming.

Residents were afflicted by compounding layers of homesickness and profound poverty was the norm.

By 2008, Black people were twice as likely as white people to still be living in their trailers (Bullard and Wright 2009: 30). The annual household income of the largest FEMA trailer park, Renaissance Village, averaged $5,000. Over 40 percent of children under four years old in that park were found to have iron deficiency anemia: a rate twice that of children in New York City's homeless shelters and "by far the highest yet documented" in the United States (Children's Health Fund 2008: 12). The study also found that 42 percent of children who visited their mobile clinic were diagnosed with allergic rhinitis and/or upper respiratory infections, and 24 percent had a cluster of upper respiratory, allergic, and dermatological diagnoses (13–14). Half of the respondents to a 2006 survey of FEMA trailer parks met the criteria for major depression. Suicide attempt rates were seventy-eight times the US baseline levels (Larrance et al. 2007).

The desperation was most acute in remote FEMA trailer parks, which also had a reputation for high rates of drug usage. Visitors to the Scenic Trails FEMA trailer park in rural Mississippi in 2007 would find trailer windows blacked out with tin foil, accusations that an unknown resident was executing others' pets with antifreeze, and an alarming number of people expressing suicidal ideation (Spiegel 2007). Located thirty minutes from the nearest town, the trailer park's inhabitants could be seen walking along the highway to work. Facing steep travel costs and pulled out of community with free family babysitting, staying unemployed was an even more reasonable financial decision for some. Shared infrastructure and illness did not always and everywhere reorient inhabitants toward collectivity. Indeed, the residents of Scenic Trails, and many other remote parks, became depressed and self-isolating in unison. Other FEMA trailer inhabitants became too sick to think about anyone but themselves. Some, of course, felt no negative impact at all.

One afternoon in August 2007, a year and a half into his trailer residency, a stranger appeared at Ray's door. She said she would like to test his trailer for formaldehyde. He wasn't sure what that meant, but he was happy to "let her do her thing." Weeks later, he was told that he had a reading of 323 parts per billion (ppb). This reading was more than three times higher than the US Environmental Protection Agency (EPA) recommended for the maximum air level for formaldehyde indoors, 100 ppb, a level that is itself more than twelve times higher than the Agency for Toxic Substances and Disease Registry (ATSDR) threshold for possible negative health consequences, 8 ppb. Having now spent years testing homes with high formaldehyde, the prospect of spending even ten minutes in an atmosphere that thick makes me queasy. "I didn't think it was a big deal," Ray shrugged and took a big slurp from his drink. "I didn't feel no side effects from it, ya know? I've always been tired." Ray thought and still thinks nothing of it. Feeling no symptoms of his own, Ray never sought out more than a cursory engagement with other FEMA trailer residents.

As time went on, most residents who had returned were able to, at least partially, buck the psychic pull of sorrow, despair, and depression. The region rebuilt slowly, and by July 2007 the majority of trailer residents had moved into permanent housing. Over time, staying in a trailer became an indicator of who faced barriers to getting back on track: largely the elderly, poor, or disabled.

When I last saw the man whom I mentioned at the beginning of this preface, Jeremy, in May 2011, he sat under the live oak tree. Two letters lay on the grass in front of him, pinned beneath his sneakers. One was from FEMA, fining him $800 for not having found permanent housing as the sixth anniversary of the hurricane approached. The other was from the city, fining him nearly $100 a day as they reinstated a pre-Katrina city ordinance banning the placement of trailers within the Orleans Parish. At this point, his family was living in one of only a handful of trailers left in the city. New Orleans was trying to move on. The last emergency hiatuses of established law were being disassembled.[8] Press coverage of the forced exodus of the last remaining FEMA trailer inhabitants was polarized. Some chided those still in trailers, labeling them lazy freeloaders past due for the boot. Others admonished the move as an antipoverty policy. All this man knew was that he didn't know where he was going to go. "It is severe. It is *severe*. It is **severe**," he repeated with increased intensity as the reality of the situation seared him in real time. After his words ran out, he continued shaking his head, anxiously running his tongue across his teeth and under his lips. The gold caps on his incisors were wearing away, revealing clean white ellipses of bone.

Before the storm, Jeremy was a pastor, with a congregation of 250. He used to rent a house in a relatively affluent suburb, and he, his wife, and their daughter were in fine health. The rubble that lay a few yards away from us, in the adjacent lot, was all that was left of his church. He and remaining churchgoers had put in new windows and a new roof, but they could not get the building back up to code after the storm. From the small window of his trailer, he watched the city's bulldozers topple his boxy, two-story church. "I almost feel like Job," he said, still seated and pawing at the grass with his clean, outdated Nikes. Much of his congregation never returned from evacuation. Any tithe collected from those who had returned went to buying school materials for the neighborhood children or keeping the lights on for elderly church members. Fire ants had crawled into my shoes and bit my toes while we talked. I let them burn for a second before walking home to get the number of a pro bono lawyer who could attempt to legally delay a forced departure. As we agreed, I texted him the name and number of a willing attorney. A month later he texted back, "They forced us out."

INTRODUCTION

Homesick, Otherwise

As the second half of the twentieth century progressed, the air within US homes became increasingly infused with a range of toxic substances. Corrosive, enfeebling, and even suffocating atmospheres became a fixture of the ordinary chemical ecologies of American housing through rapid and large-scale shifts in building practices that took place beneath the surface of a home's architecture.

The chemical reactions that formed the first plastic—phenol and form-aldehyde in Bakelite—were no longer reserved for items like jewelry, billiard balls, and the field telephones used in trench warfare but began to clad houses as the adhesive that held together plywood and particleboard. The war effort had accelerated the development of these engineered woods that utilized petrochemical binding agents, replacing animal blood and plant-based glues. These standardized sheets of wood-plastic mixtures resisted decay, cut and deskilled installation labor, and reduced the number of nails used in average home construction by half. From the early 1950s to the 1990s, the average square footage of plywood used in home construction quintupled. The use of particleboard, another engineered wood product, increased by 3,000 percent (Ore 2011: 270, 272).

Urea-formaldehyde resins progressively took hold in interior applications, spanning plywood, particleboard, chipboard, Luan, and a whole host of other engineered woods, because they were less expensive than phenol resins. Phenol-bound boards were reserved for exterior sheathing, where weather conditions demanded their ability to withstand moisture. But cheaper urea-formaldehyde chemical bonds are more unstable, and, importantly for our story, they continuously and quietly sublimate formaldehyde into the nation's most intimate breathing spaces. Older toxic domestic building materials would need to be damaged—peeling lead paint or an errant saw blade cutting into asbestos adhesives—to enter human bodies. Formaldehyde needs no such rare invitation to spring into the air. It constantly and invisibly wafts from solid form into inhabitants' lungs.

As more toxic chemicals sloughed off building materials into interior air, the rise of modern climate-control systems worked to seal those emissions inside the home (cf. Barber 2023). The 1970s oil crisis then accelerated this trend by giving rise to federal policy and state building codes that aimed to more tightly partition homes from the outdoor environment in the name of energy conservation (Jacobsen and Kotchen 2010; Wolfson 2013). By 1995, one in four homes had poor ventilation (Miller and McMullin 2014).

Today, an invisible mist of microemissions cascades in slow motion from the adhesives in your hardboard kitchen cabinets; wafts up from your carpet, laminated flooring, or particleboard subfloor; and swirls in unseen vortices around your passing ankles. Without a cracked window, an opened door, or other forms of air exchange, these chemicals accrue, molecule by molecule, within the building envelope of your home. Nine out of every ten breaths in your life will take place in enclosed spaces where the air is suffused with chemicals that are off-gassing from construction materials and the myriad commodities that fill our daily lives. Indeed, the average levels of formaldehyde found in conventional American homes are two to three times higher than the level the EPA says can decrease pulmonary function and aggravate asthma and allergies (US Department of Housing and Urban Development 2024; US Environmental Protection Agency 2024).

The dual dynamics of more chemicals emanating from building materials and more tightly sealed homes bottling up that toxic brew have yielded indoor toxicant levels often many times higher than corresponding outdoor air (US Environmental Protection Agency 1987; Khoder et al. 2000). This is where America lives: indoors, where the errant molecules of the built environment's extended exhale are met by the inhales of life. Here, within these heightened spaces of exposure, North Americans, over the course of a day, gulp some eighteen pounds of industrially textured air. Respiration is the most prominent mode of exchange between human bodies and the rest of the environment (far greater than eating or drinking), and it is a primary site for absorbing the residues of engineered materials.

US residents are very literally homesick within their own homes, and chemical exposure is only part of the story. This book breaks this homesickness down into three forms: toxic homes (the subject), the sick concepts and markets of racial capitalism (the condition), and a homesickness in search of more just futures (the method)—a discontent with and longing for something better than now.

Toxic Homes

Ask the question of how shelter became exposure, of how the average American home became an epicenter of chemical hazard, and you will find an answer a mile below the earth's surface.

Formaldehyde, the most common and most toxicologically understood indoor-air pollutant (Salthammer et al. 2010: 2537), is made from methane, also known as natural gas. Methane is an inexpensive and relatively abundant raw material for producing a wide variety of chemicals, also known as a feedstock. The scale, profitability, and low cost of natural gas is primarily driven by its widespread use in generating electricity and heating homes. Further financing its low cost, natural gas is sometimes considered a "freebie," a fuel extracted by convenience since it perches atop more valuable oil deposits. Chemical exposures within the home, in other words, are driven by a world that runs on hydrocarbons. Vast societal investments in fossil fuel consumption and extraction subsidize the affordability of the toxic home and complicate the economics of building differently. The planetary homesickness of climate change and the individual gassing of people in their homes are not different issues.

Beyond being derived from a greenhouse gas, formaldehyde—that binding agent that holds together contemporary homes—is an irritant, an allergen, a neurotoxin, and a known human carcinogen. The chemical causes more cancer in humans than any other known hazardous air pollutant, by far (US Environmental Protection Agency 2025). Its presence in mammalian bodies can destroy enzymes that maintain bronchial tone, strip axons of their sheathing, dysregulate gene expression, break chromosomes, misfold proteins, exacerbate asthma, degrade semen quality, and impair cognition and learning (NTP 2011; H. Wang et al. 2015). Neurochemists increasingly suspect that this nearly omnipresent chemical has a role in neurodegenerative conditions such as Alzheimer's disease, accelerating the aging of our minds as it gives form to our homes (Jing et al. 2013).

But this story is more complicated. Formaldehyde is essential not only for the building techniques propagated by industrial capitalism, but also for life itself. Formaldehyde predates carbon-based life forms on earth and likely played an important role in the origins of life in primeval oceans. In the human body, the chemical is an indispensable metabolic intermediary in the biosynthesis of two of the four building blocks of DNA, some amino acids, and molecules that play a role in blood pressure control and hormone signaling (NTP 2011). At any given moment your blood carries detectable levels of formaldehyde. Your brain likely hosts two to four times the level present in your blood (Tulpule and Dringen 2013). Chemical industry advocacy groups are fond of these facts and assert that the average human body essentially takes a shot (1.5 ounces) of self-produced formaldehyde every day (American Chemistry Council 2024). This internal dose is also always causing DNA damage (Leng et al. 2019). The role of formaldehyde cannot be labeled as strictly physiological or pathological.

Formaldehyde protects home construction materials from insects, bacteria, and fungal decay. It reduces the costs of construction while also hastening the decay of human inhabitants. The irony is that this chemical attacks human tissues but also fends off postmortem decomposition, as gallons of it are injected into the veins of American corpses as an embalming fluid. The compound chemically tugs people toward death and then defends their inanimate bodies against biological disintegration. Take a step back and the multiple, oppositional roles that formaldehyde occupies are also those of the engineered home: an agent of preservation that is also an agent of decay, a shelter that is the seat of exposure.

Thinking more broadly than this single chemical, domestic air—meaning the air within a home—as a whole is unregulated in the United States. Global maps of domestic air-quality regulations gray out the country or lay a large question mark over it. The only federal domestic air-quality regulation in US history that I'm aware of was announced in April 2008, only to be quietly rescinded some three years later. It only applied to one specific kind of home— emergency housing units made by the Federal Emergency Management Agency (FEMA)—and only applied to one kind of chemical: formaldehyde.

Something happened in the FEMA trailers that broke with how regulators understood chronic chemical exposures in the home. The more than 120,000 government-issued temporary homes that provided refuge to Gulf Coast residents allowed these invisible chemical hazards to be uniquely concentrated, documented, and newsworthy. Indeed, the FEMA trailers arguably constituted the largest formaldehyde exposure in the history of our species, an exposure event that took place in the media maelstrom that followed the costliest natural disaster in US history. The trailers and their toxic air garnered congressional hearings and even drove the first set of revisions to the backbone federal toxics legislation, the Toxic Substances Control Act, focusing on formaldehyde.

But the FEMA trailer is not an aberrant shelter, a mere upshot of postdisaster disarray. Rather, it is an indicator of the commodity pathways, dreamworlds, and legal structures of the home that, today, can be described by no word other than conventional. These trailers are inroads to all of our homes. For this reason both formaldehyde and the FEMA trailers anchor the subject matter of this book as it attempts to grapple with much more widespread and uneventful domestic health hazards.[1]

Homesick for Otherwise Futures

I approach toxic homes with a method that is homesick for a future that does not look like today. By that, I mean this book is committed to advancing, documenting, and theorizing horizons of change. An "otherwise" is shorthand for the destination of various change-making actions. The method of this book involves learning from chemically exposed people about how they change the microconditions of life to enhance their survivability. It also involves throwing my weight—always with others—against the larger forces that drive mass corrosive gassing. This book also imagines how to destabilize the global hydrocarbon extraction system that underwrites a world built by formaldehyde. This form of homesickness is not fueled by a romantic nostalgia for a pristine and prior home. Instead, it is a homesickness for something one might feel is missing, but has yet to experience—a simple conviction that the state of things is not right.

Hungry for minuscule and massive destabilization of the violent status quo, the book documents my collaborative work in the service of making this future: poring over evidence in toxic tort litigation as a law clerk, designing open-source and low-cost tools for impacted communities to monitor and remediate their domestic exposures, creating air-monitoring protocols with communities living in the fracking fields, building remediation systems, producing a mini-documentary, and collaborating with artists and engineers to envision how needs and desires can be met without advancing solutions that shape-shift or spatially redistribute the harms that are tied up in the petrochemical home. I stress the collaborative nature of all this work because a commitment to co-learning—muddling together rather than sitting in analytical judgment—is foundational to all of these projects.

A homesick longing for a different world, a more just world, drove a decade of work that was focused not just on narrating the plight I encountered or making sense of what I saw through theory, but also on putting to the test dominant ideas of how to improve the conditions of life. As a result, a central focus of the book is the actions that people afflicted by environmental injustices (and their allies) are told will better their situation. I'll refer to the logics behind common answers to the question "What can be done?" as theorizing change:

reducing harm, documenting harm, changing systems, and so on. Theories of change are often tacit, guiding how groups come together to improve their environments or fragment as they disagree what kinds of interventions are best suited to any particular issue.

Grounding my method in theorizing change roots the book in accountability to the situations and people with whom I became involved in this research. At heart, theorizing change represents the desire to have a plan to interrupt or eradicate harm.[2] Theory in general, on the other hand, is primarily accountable to explanation and colleagues within scholarly discourse. Of course, theory is crucial to and informs theorizing change. But, in the case of this book, simply making sense of what I observed through theory left me with the feeling that vital work was left undone. This undone labor was jointly ethical—how can my research be put to work for the people who shared their stories with me—and analytical—what do different interventions reveal about power and the resilience of harmful systems, concepts, or practices?

All research that is not content with the status quo is at least tacitly theorizing change (see chapter 5; Tuck 2009). Pulling these unspoken theories into the analytical limelight allows me to interrogate them and make my inquiry more robust. Extending ethnographic analysis to my own humble attempts at bending the future away from harm helps me foster accountability with a discipline, anthropology, where spurring meaningful and substantive change is not the essential question.[3]

In this book I attempt to hold, in a sometimes quarrelsome and often wary alliance, multiple forms of stopgap, reformist, and radical change projects aimed at ameliorating or preempting homesickness. Just as those exposed to toxicants today cannot be held hostage to quasi-plausible massive systems changes of an unknown tomorrow, we are consigning the world to perpetual toxicity, and those in struggle to perpetual labor, if prevailing technocratic triage continues to dominate how environmental improvement is imagined.[4]

These methods are a far cry from simply reaffirming "a cosmic faith" (Haraway 2016: 3) in the ability of technology to solve social problems. I am also not interested in simply pointing to the many instances of tragedy or farce that are strewn across the paths to environmental justice. Each way of theorizing change has a logic—the asserted mechanism of transformation—and a temporality—the timeline for it to take hold. This book textures the multiple, and often mutually exclusive, logics and temporalities of intervention.

Answering in resounding affirmation to María Puig de la Bellacasa's question "Couldn't holding these apparently contradictory positions together open unexpected avenues to think environmentalism?" (2017: 128), I advance a syncretic method rife with productive frictions, one that makes strides toward a less toxic world by coordinating across multiple, self-enclosing timescales and logics. This method is not a hopeful metaphor, a conjectural agenda, or a feel-good holism but a collection of practices, abounding in internal tensions, that I have been co-learning in constant relation with my theoretical and empirical work and with dozens of collaborators.

OF ALL OF THE HUNDREDS of moments that set this method into motion, one in particular pushed me to reckon with an assumption common to almost all of the conventional theories of environmental change. As I was following formaldehyde up the production pipeline to the extraction of natural gas, from which it is derived, I found myself seated next to two white residents in a rural community that had been gripped by intensive fracking for six years. Form-aldehyde levels of 30 to 240 times background levels had been found wafting through the air of their formerly agricultural community (Macey et al. 2014).

I pulled over a chair to the kitchen table as Frank popped three frozen burri-tos into the toaster oven. Frank often referred to his ten-acre plot as "the dough-nut hole." He was the only one in the area who hadn't leased their mineral rights for natural gas exploitation. An active well pad stood just on the other side of a small hill in his backyard. Frank's antifracking activism is legendary. He spent his blue-collar life savings buying expensive environmental monitoring equip-ment and driving the country in an old RV, helping communities document their exposures from oil, gas, and petrochemical infrastructure.

Rachel, who lived about twenty minutes away, sat on the other side of the table. She was the outspoken leader of the local group of residents concerned about the impacts of fracking. As we spoke about the group's evolving strategy she leaned in and explained, "There's been a lot of lengthy articles about [our local environmental justice community group]." She spoke with a surprising air of discontent. I had assumed that sustained media attention would be valuable leverage for their mission to foster accountability in their industrial neighbors. Reading the confusion on my face, she elaborated, "It was good but, um . . . yeah. We are getting slammed by people all over the country that don't want fracking to happen because. . . ."

After a pause, Frank attempted to finish the thought that had evaporated from Rachel's mouth. "Their way of thinking is that if we make it less bad. . . ." Frank too seemed incapable of completing their shared impression of fellow environmentalists with whom they'd struggled arm in arm for years. One of Frank's hands alternated between ruffling and then smoothing his graying beard as he looked past me. The tick of the toaster oven made the passage of time audible. Rachel then succinctly distilled their reluctant view: "They want it to get as bad as possible here."

Rachel, Frank, and their local allies had already lost the local battle to prevent fracking for natural gas. After early spectacular contamination events (like drinking water coming out of the tap like a flamethrower) failed to halt the rapid industrialization of their rural county, they pivoted their organizing efforts. Only denouncing what had already arrived felt futile, so they sought survival by brokering an often deeply technical dialogue with industry about mitigating air-quality impacts. In other words, when it became clear fracking could no longer be prevented or stopped, they shifted to trying to make it more conducive to staying alive. In one of my first meetings with them, they were working on introducing a company that makes high-efficiency catalytic converters (a machine that scrubs pollution from emissions) to the biggest oil and gas company that was continuously releasing a slew of toxics into their region.

Parts of this bucolic community had worse air quality than London, a city of over eight million people where I was living at the time. A constellation of polluting infrastructure owned by different companies netted the rural landscape. Concerned residents like Frank and Rachel wanted to attribute the clouds of pollution that would accumulate in different valleys to specific companies in order to trigger a regulatory violation. But this was both scientifically arduous and required vast amounts of data that was well beyond the motley group's resources to collect. Governmental oversight felt as useful as catching emissions in butterfly nets. In the wake of a failure of regulation, the work of persisting took the shape of liaising between corporate entities in the hope that these profit-based industries would implement safety measures in excess of the letter of the law. The group continued to document harm in the hope that it would facilitate their mitigation negotiations.

As a result, the wider antifracking community accused them of flipping. "We've been the poster kids for them," Rachel almost wistfully remembered. Indeed, many of the most iconic fracking-related contamination events had befallen their corner of the Marcellus shale field (explosions, corrosions, and a slew of reported human and animal health issues). In the middle of our work together, Rachel's well, the source of drinking water at the home where she cared for an elderly father, was contaminated by fracking chemicals. Documentation of these happenings was valuable currency for the larger antifracking movement; mitigating these harms could be seen as making the process more palatable, facilitating its entrenchment and expansion.

As we talked, headlights from an industry truck cut through Frank's forested backyard. It was turning off the paved highway and down a dirt road to a nearby well pad that hosted around-the-clock activity. He and his late wife, who had recently passed from cancer, built their house in its entirety. Its only embellishments are the four tiny dovetail joints that Frank, a retired cabinetmaker, had chiseled into every floorboard—holding them together with excessive unity. Frank stood up and moved silently over the creakless floor toward the fridge. Grabbing beers for all of us, he concluded decisively, "If we don't die, it's bad for them."

What was nearly unspeakable at Frank's kitchen table—that the sacrifice zones integral to maintaining industrial production can mirror sacrifice zones for environmentalists seeking evidence of harm to oppose those very same industries—is just one of the tensions knotted into that somber exchange. One could read the strain of these circumstances as the double binds—situations of irreconcilable choices—of contestation: as struggles between the needs of hyperlocal survival and a more generalized future or between reform and total refusal.[5] But what strikes me as more important than the friction of divergent logics of change is what they have in common.

Those hoping to abolish fracking and those whose lungs and drinking water are marked by having lost the local struggle to keep fracking out have a *shared* dream that documenting harms will lead to the realization of their divergent goals. The reason I was seated at that table was to monitor the air impacts of nearby fracking infrastructure through an imminent multiyear community air-quality monitoring project with Frank, Rachel, and their neighbors.

Whether by attempting to accumulate air-quality data that would force local compressor stations (essentially small refineries that are part of natural gas transportation infrastructure) to reduce air emissions, or by documenting human illnesses in pursuit of stopping and reversing fracking as a practice, both strategies see evidence of risk or harm as key to change. More than this, the groups implementing these strategies are tirelessly laboring to accrue *enough* evidence. The idea is to rupture "the barrier of public indifference [. . . by] marshalling all the facts" (Rachel Carson quoted in Brooks 1989: 258), by making the evidence substantive enough, eventful enough.

Enough. An imagined threshold of eventfulness leads to change. But can events make good on their promises of change? Or does working to collect perpetually almost-sufficient data partially function as busy work, ultimately guaranteeing that things remain the same?

The diverse array of conventional avenues for detoxification largely buy into, fall prey to, or even simply acquiesce to what might be called the eventfulness myth. The eventfulness myth is the promissory note we write to ourselves and each other when we imagine that we can document fastidiously *enough* or narrate charismatically *enough* that we can extinguish a sufficient amount of the environmental problems of late industrial capitalism. Other times, we anxiously anticipate things to eventually go wrong *enough*—for factory explosions to be destructive enough, or, in related social justice endeavors, for markets to crash deeply enough, or wrongheaded policies to fail egregiously enough.[6] *Enough is enough* as a rallying cry at protests is sentimentally true, but empirically it may be more the exception than the rule.

In the case of toxics, the lack of eventful events means that slow, agentless exposures, with their ongoing subclinical illnesses, are less likely to inspire public and scientific concern. Such a logic, the very logic of the event, also cultivates an imagination of a threshold that stimulates change once attained, one that turns out to be an ever-receding horizon.[7] The specter of the event defers substantive change to the catastrophes and representational technologies of a tomorrow that never comes. As Joseph Masco has articulated from a slightly different vantage, the idea of a "crisis" built on an accumulation of eventfulness "has become a counterrevolutionary force in the twenty-first century, a call to confront collective endangerment that instead increasingly articulates the very limits of the political" (2017: S67).

As the writing of Saidiya Hartman has made clear, what can feel like powerfully bearing witness to the evidence itself—the stories, images, or statistics—can quickly slide into an immured spectatorship of "endless recitations of the ghastly and terrible" (1997: 3). The empathy fuel of change ignited by an event, crisis, or "the exposure of the violated body" is particularly ineffective, or at least fickle, in struggles to reckon with the violence of structural discrimination (2–3). Organizer and public intellectual Mariame Kaba has labeled attempts to leverage extraordinary violent events in service of substantive changes to the everyday as "a fool's errand." This approach often simply does not work. In the case of anti-Black state violence, particularly, "it leaves Black people in the position of having to ratchet up the excess to get anyone to care or pay attention" (Kaba 2021: 85).

Far from spurning the meticulous fact-finding toils of others, the bulk of both this book and my own practices falls within the frame of the standard, yet ever-shifting and slippery, theories of change. Coming to terms with the likelihood that there will never be enough data, that many science-for-justice endeavors are running on a "data treadmill" (Shapiro et al. 2017), producing ever more data toward an intrinsically unreachable destination, was utterly bleak. Mourning the dramatically constrained horizons of a familiar hope initially led me to an extremely cramped headspace. The possibilities this critique opened up were so alien to my corner of the toxics activism world that they approached logical heresy. But soon, after a melancholy year I spent studying, experimenting, accepting serendipitous invitations, and soliciting guidance, the claustrophobic outlook I had cultivated in myself though this critique of event and evidence opened out onto an overwhelming expanse of alter-political possibility.

In their work, historian-organizer M Murphy (2004, 2006, 2007, 2008a, 2008b, 2010, 2013, 2016a, 2016b, 2017) has documented how technoscience dilates our ability to make claims about systemic and unequal environmental injuries at the same time as it constricts how we can authoritatively know our environment and how we imagine chemical relations starting or stopping. This paradox leads them to simultaneously search for accountability within flawed systems of environmental science and regulation while also attempting to articulate the harms of late industrial worlds beyond the limited way that industrial chemistry knows its own problems. In other words, Murphy works both "with and against technoscience" (2017: 495) in their attempts to bend asymmetrically benefiting and burdening chemical infrastructures toward justice and equity.

In this book, I hope to detail how the axis of conspiring and condemning has more degrees of freedom, more political orientations, that cannot be reconciled with the adversarial logics of antipolitics.[8] Murphy's complex political posture opens the possibility of alter-political action. More succinctly, not only "with and against" but "with and against and of and beyond" this world. "Of and beyond" is an orientation undeniably anchored in material realities but is oriented toward an otherwise and not only toward one that disavows in antipolitical action.

Apprehending how environmental reformism and even righteous means of contestation often lead to entrenchment rather than resolution pushes me, not to spurn survival work like that of Frank and Rachel (and much of this book), but to supplement ongoing work aimed at either mollifying or dismantling our engineered world, bolt by bolt. Highlighting how the hegemony of adversarial, event-based modes of change uphold ostensibly compulsory harms enables other visions of change to begin to appear reasonable. The liberatory value of enticing abandonment of the status quo, rather than exclusively locking horns with it, is increasingly clear. From this perspective, it's easier to view the multiple qualities of direct contestation as a necessity at some timescales and as attritional and conservative at others if it's not allied with alternatives. What falls out of reason in thinking with and against and of and beyond the current world is the monolithic dominance of the logic that "necessary" harms are indeed necessary, only to be forever mitigated and never upended (Weizman 2011).

My goals with all of this big talk are simple. To better understand what the conventional tools for detoxification—litigation, legislation, remediation, and enumeration—can and can't do and to ask how alter-political action can be a vital addition to the toolbox. We arrive at these broad ends with cross-cutting applicability by way of the specifics of the toxic homes of the FEMA trailers.

Finding Home Under Racial Capitalism

As the city and the federal government were forcefully closing trailer doors as federal property for those displaced by hurricanes with fines for remaining low-income residents in 2010, trailer doors were reopening all across the United States as private property for those displaced by the ordinary functioning of capitalism. Mediated by rural speculative entrepreneurs and the open market, FEMA trailers were becoming new homes across the country.

By that point, returned or unused trailers had stood empty for years. A November 2007 court order had forbidden the US government from selling used FEMA trailers until January 1, 2010, retaining them as possible evidence in a lawsuit related to formaldehyde exposure. After the moratorium ended, the federal government began reselling some 150,000 of these now-infamous trailers at public auctions and at fire-sale prices.[9] The General Services Administration (GSA) was tasked with offloading these trailers in massive auctions. They attempted to sever the health liability of selling potentially contaminated units by adhering stickers on the windows of the FEMA trailers that stated, in red lettering, "NOT TO BE USED AS HOUSING." Many of the units I tracked would indeed be used and resold as housing. Much more often than not these warnings were removed.

The vast majority of the people I talked to who lived in resold FEMA trailers across the country were low-income and white. The boxy, 250-square-foot homes link the organized neglect of Black people in New Orleans after Katrina to the mass toxic exposure of low-income white people. The disjointedly shared space of the FEMA trailer connects a singular urban horror, which some think of as vital fuel for what would become the Black Lives Matter movement (Bouie 2015), with white rural or peri-urban spaces that many in the United States assume to be bastions of racial animus. Simultaneously, at least a thousand FEMA mobile homes were donated to tribal governments hard struck for housing (FEMA 2007). The story of these trailers is a story of racial capitalism. A story of anti-Blackness, settler colonialism, and a white underclass that is not claimed by whiteness. A story about the distribution of an acute exposure within a form of capitalism that tends "not to homogenize but to differentiate" (Robinson 2000: 26) and to extract value from those social differences. A story of temporary and long-simmering capital and housing crises, of disaster capitalism melting away to reveal more continuous contours of capitalism and the state.

Racial capitalism, which is the idea that racism precedes capitalism and persists as a structuring logic, is the condition of possibility for this distribution of toxic living. In *Black Marxism: The Making of the Black Radical Tradition*, Cedric Robinson argues that racial capitalism emerges not from early colonial encounters between Europeans and non-Europeans but from intra-European dynamics. Economic exploitation and hostility toward, for instance, the Slavs and later the Irish were justified along racial/nationalistic lines. This racial differentiation was then baked into capitalism, forming "its epistemology, its ordering principle, its organizing structure, its moral authority, its economy of justice, commerce, power" (2000: xxxi). These ideas and practices were later extended to and intensified upon other racialized communities across the globe. Value derived from racial difference was baked into the very core of the US economy in Black chattel slavery and the quest of Indigenous elimination and continues to shape landscape, livelihoods, and longevity (Pasternak 2020; Estes et. al 2021; Jenkins and Leroy 2021). Geographer Laura Pulido (2016) argues that a lack of reckoning with racial capitalism is a key reason that environmental injustices have not diminished, and may have worsened, in the forty years since the inception of the environmental justice movement. I agree.

Following an environmental hazard from a predominantly Black community to a predominantly white population is not a well-worn path, so let me clarify the group of white people who inhabit these trailers. White people living in mobile homes are often referred to as "trailer trash," a term closely allied with the larger umbrella term of "white trash." In the late twentieth century, living in manufactured housing developed a stigma. The number of postagrarian rural communities increased steeply during this period (Castle 1995), decoupling rural living from the institution of farming. This shifted the rubrics of rural social status from an emphasis on family reputation to that of residence location and conspicuous consumption. At this time the mobile-home park became a significant rural community form and non-trailer-dwelling whites began wielding terms like "trailer trash" to erect a racial firewall against the other whites who resided in these marginal homes. "White trash" became "feared and despised because of their economic and physical proximity to minorities" (Kusenbach 2009: 402).[10] In the latter half of the twentieth century, the trailer became symbolic of poor whites, whom generations of eugenicists had described as shiftless, bad breeders, living hand-to-mouth, and unwilling to work in the first half of the century (Isenberg 2016: 315), all derogatory labels that eugenicists and public authorities aimed more squarely at Black and Indigenous populations.

This is not to say that white people spurned by whiteness can't also be anti-Black. I remember talking one day with an older white man in a trailer park just east of New Orleans. He spoke to me though a gap in the slats of his trailer window blinds, held open by his fingers. He denied formaldehyde's existence ("That [formaldehyde issue] is just a bunch of nonsense made up by n****rs to get a buck") even while admitting that the formaldehyde, which he claimed he could smell, was helpful to him by keeping vermin out. He also refused to believe that his compounding health issues had any relation to his home. Moments like these, which closely resemble the phenomenon that Jonathan Metzl (2020) describes as "dying of whiteness," in which white people and their racism are on the hook for their ill health, are a minor current in this story. Rather, the dominant current in the trailer diaspora is down-and-out white folks being subject to modes of structural marginalization intended for and prototyped on Black and Indigenous people (cf. Pulido 2016: 2). Racial capitalism does not care about white people, only whiteness, yielding opportunities for differentiation and value extraction.

The majority of mobile-home purchases are financed via personal property loans, similar to those used for cars. These loans provide greater returns for lenders, as they bear higher interest rates than mortgages and come with fewer consumer protections. These loans also carry fewer up-front costs. As time goes on, this yawning gap between real estate and "wheel estate" expands as mobile homes tend to decrease in value—the opposite of conventional homes.[11]

Our climatically destabilizing planet will only compound these problems. The walls of mobile homes have limited integrity (with limited insulation and poor seals), making the units difficult to heat and cool and saddling residents with exorbitant energy costs. Mobile-home residents are twice as likely to bear a high energy burden than residents of a single-family detached home (Berry et al. 2018). A just energy transition is further away for mobile-home residents than for any other US housing type, with a net return on energy efficiency investment twenty-four times lower than in single-family homes (Tonn et al. 2018). These costs are not just financial. Mobile homes are increasingly placed in the hottest census tracts and on floodplains (Lee and Jung 2014). In Arizona's sprawling and sweltering Maricopa County, for example, 28 percent of indoor heat-associated deaths occur in mobile homes, despite only amounting to 5 percent of the housing stock (Kear et al. 2020).

The persistence of housing insecurity, which often makes these homes the best option for those priced out of the conventional housing market, makes mobile-home parks lucrative investments for global capital. For this reason, private equity firms, ranging from Blackstone ($457 billion in assets) and Apollo Global ($270 billion) to the sovereign wealth fund of the Government of Singapore ($360 billion) and the Pennsylvania Public School Employees Retirement System ($56 billion), have converged on purchasing manufactured housing parks since the mid- to late 2010s (Baker et al. 2019). The investment value of mobile-home parks increases during recessions. Rents and fees in these parks are on the rise, forcing some people on fixed incomes to skip doses of prescribed medicines to stretch out the time between refills.

Together, these attributes make it easy for residents to fall behind on payments. Repossession is swift. Defaulting owners sometimes return home not to foreclosure notices or a changed lock, but to the negative space of their home having been towed away to an impound lot with their belongings inside it. Repossessed mobile homes are then sold for prices near the cost of a new home, but with a lower down payment and vastly reduced warranties.

White people are more likely to live in mobile homes than Black and Latino people, and they are especially likely to turn to mobile homes for retirement (Wilkinson 2020). While communities of color are less likely to inhabit these homes, when they do, the financial cuts are deeper. Mobile homes and manufactured housing have been key technologies of settler expansion, colonialism, and financialized extraction from Indigenous people in what has become known as North America since whites entered the continent. The first manufactured housing on the continent (unless you think of boats as ancestors of the mobile home) dates back to 1587. Manufactured houses proliferated across European colonial empires, like thousands of flags staked in the ground. Prefabrication of framing and facades allowed European colonists to instantly reproduce the aesthetics of home; seize territory through European legal doctrines like terra nullius; advance ideas of universal white superiority; protect and advance regimes of private property; actively remain ignorant of local materials, local ecologies, and place-based practices of homebuilding; and rapidly scale settlement. Mobile homes enabled the mobility of imperialists (Shapiro 2019a).

Settler colonialism is driven by one form of homesickness and manifests as another. Colonizers set sail suspended between an imagined future home, for which they searched, and the compulsive urge to re-create the home that they left behind (cf. Van Herk 1991). The imposition of that dream/memory on Indigenous or otherwise colonized peoples through land theft, slavery, and extreme, protracted violence wrenches them too into a homesickness of a very different kind.

Today, the mobile home's relationship to colonialism is not just a vessel for fast, affordable settlement. Rather, Native Americans on reservations live in mobile homes more than any other social group. For example, in 2016, American Indians and Alaska Natives living on reservation land were twenty-one times more likely to finance a mobile-home loan versus a conventional home loan than the average American (Johnson and Todd 2017). The omnipresence of trailers on reservations stems from some two hundred years of the United States failing to uphold treaty obligations to Native American tribes stipulating that the United States would supply housing, not out of any form of charity but as a part of violence-backed exchanges for land (V. Davis 2002).[12] Trailers fall into the housing void of two centuries of double-dealing.

The United States holds most Native land in federal trust. This forced separation of land and homeownership aligns with what some call the "halfway homeownership" (Sullivan 2014) of the mobile homes that are spread across Native lands. In the early 1960s, the Department of Housing and Urban Development (HUD) began providing some funds for Indigenous housing predominantly through manufactured homes. "Hudda" has become shorthand on reservations for the boxy, manufactured homes that dot these sovereign territories (Edmunds et al. 2013). With plywood walls, fiberboard doors, laminate in-built furniture, and particleboard subflooring, no housing type uses as much polymerized wood as mobile homes.

Prior to HUD more closely regulating the construction of mobile homes in 1976, the formaldehyde levels in these state-provided homes were likely unimaginably high. By the twenty-first century, upward of 75 percent of these homes had become contaminated by mold due to water intrusion, climatically inappropriate design, or shoddy plumbing (Seltenrich 2012). As formaldehyde is hydrophilic, greater amounts of formaldehyde are released from these waterlogged engineered woods (Maddalena et al. 2009). Reliance on less expensive mobile homes can be further explained by the high rate of rejection for on-reservation conventional home loans (three in four applications), and, when granted, these loans carry the highest interest rates for any racial group (Cattaneo and Feir 2019).

Given the history of active neglect by the United States of its treaty obligations, it is perhaps unsurprising that the US government transferred 1,300 FEMA mobile homes to eighty-eight tribes between 2007 and 2009 (Cooper

2011). I've been contacted by members of nearly a dozen tribes about formalde-hyde testing.[13] Native Nations became a prime location in which to disappear these toxic assets from the market while appearing charitable.[14] Race- and place-based exclusions from capital had worked to establish such acute housing desperation that these toxic homes, with their limited insulation, appeared at-tractive to tribes even in significantly colder climates.

Building homes in a factory, instead of on-site, combined with racist lend-ing practices, enables mobile-home manufacturers to hyperconcentrate their profits. For example, multinational conglomerate holding company Berkshire Hathaway owns Clayton Homes, the largest mobile-home manufacturer in the United States and, by extension, the largest homebuilder of any type in the coun-try. Clayton Homes operates a network of more than 1,600 dealerships and sells almost half of all manufactured housing in the United States. Its subsid-iary lender, Vanderbilt Mortgage, has disproportionately lent to Latinx, Black, and Native mobile-home owners, at double the frequency of white purchasers (Baker and Wagner 2015).

The contours of racial capitalism are unambiguously distinct in the patterns of mobile-home financing. On average, communities of color pay an interest rate that is 0.73 higher than that paid by whites. Vanderbilt Mortgage agents have been recorded asserting to potential buyers that they are the only lender that will lend for mobile homes on the reservation, despite it being illegal to make such a claim; the tribe itself lends at a rate that is often half of what Vander-bilt charges (Baker and Wagner 2015). Similarly, the company posts Spanish-language advertisements in Texas stressing that Social Security numbers aren't needed to secure loans, then saddles undocumented immigrant families with loans that they will never reasonably be able to pay back. These impossible loans lend themselves to other forms of dispossession. On the Navajo reservation alone, Vanderbilt Mortgage had seized at least 691 mobile homes by 2015. Land that had been Black-owned since Reconstruction and "put up" to secure a predatory loan has also been repossessed (Baker and Wagner 2015).[15] From the oldest residents to its newest, hawking the American Dream of homeown-ership is infused with homesickness.

Historian Keeanga-Yamahtta Taylor (2019) has detailed how banks, white property owners, and real estate agents have greatly profited off a conventional mortgage-financing system based on financial risk assessments that cut along racial lines. This structured market drove up the value of homes in white neigh-borhoods and aimed extortionary interest rates at communities of color that were deemed riskier. In a process that is now being repeated in the mobile-home market, lenders expedited default by charging communities of color higher monthly payments for lower-quality housing than their white counter-parts. Foreclosure would then result in Black homeowners losing their deposit, their investments in improvements, and accumulated equity. Taylor incisively calls this phenomenon "predatory inclusion."

The demarcation between true affordability and predatory inclusion in the mobile-home market is deeply murky. As the largest form of unsubsidized affordable US housing—costing upward of 50 percent less per square foot than conventional homes—these factory-built homes no doubt keep roofs over the heads of tens of millions of people who might otherwise be even more housing insecure. Yet, thinking beyond their up-front cost, mobile homes swap the massive and singular weight of a large mortgage for a constellation of smaller hidden weights (chemical exposure, fewer consumer protections, mold exposure, energy poverty). Mobile-home owners who park their unit in mobile-home parks additionally face monthly lot fees, a disproportionate lack of water-service reliability (Pierce and Jimenez 2015), greater proximity to outdoor environmental exposures (Sullivan et al. 2021), and a heightened threat of eviction (Sullivan 2018). These issues totalize into a daunting set of obstacles for their low-income residents.

Riding in the back of a maintenance van in the southern Appalachians in late 2022, I caught a glimpse of how the split between home and real estate that creates the opportunity for profit to be extracted from housing vulnerability also gives rise to radical alternatives. Rather than simply lobbying for greater regulation of mobile-home tenure, financing, and manufacture, a group of predominantly Latinx immigrants had pooled their resources and bought their own mobile-home park, collectivizing landownership while keeping a private place of their own. What began as just a three-trailer cooperative has grown into a network of four parks and over eighty trailers in just a few years. These multiracial, low-income co-owners came together via shared need, even if they didn't share politics. Other cooperative businesses (translation, child care, real estate, bookkeeping, property maintenance) have sprung up alongside the housing co-op. The maintenance co-op helps strip old, sometimes abandoned trailers down to rebuild them with the materials of a conventional home (see fig. I.1). This doesn't make them toxicant free, but it vastly reduces the heightened toxic burden of manufactured housing and any older housing that has significant maintenance issues.

On the wintry afternoon that I was visiting the group, part of the leadership was visiting a new property for which the financing had come through hours before via a national community wealth cooperative called Seed Commons. A group of five adults and an equal number of kids hooted and laughed as they

FIGURE 1.1. Photo of a mobile-home interior under renovation with hardwood floors and interior walls that are not made of engineered woods, December 2022. Photo by Nicholas Shapiro.

imagined all the happenings that would fill the cavernous building, slated to be an event space, store, and offices. Amid sharing their successes, the leaders made it clear that the acuteness of their members' housing vulnerability—immigrants having homes titled to citizens that could "get sold out from under them" and with no access to financing in their own names—drove their need for experimentation. This group, which works with a constellation of cooperatives under the umbrella of Colaborativa La Milpa, is not alone in banding together in this way. Approximately a thousand mobile-home parks across the country are now resident-owned communities. Many date back to the 1950s and 1960s and have a more conventional homeowners' association nonprofit at their center, making the community I visited on the outskirts of Asheville, North Carolina, rather unique.

Homeownership constitutes both the major driver of family wealth in the United States and the major driver of racial inequities in wealth. Even as homeownership inequity decreases, communities of color accumulate equity more slowly than white communities and have lower home values (Cattaneo and Feir 2019). And, as Taylor (2019) has argued, lending companies have extracted additional profit from Black homeowners' desire to participate in wealth generation. The trailer parks I visited in North Carolina taught me another way forward. Land can be quickly collectivized, rather than better marketized, through the separation of real and wheel estate. With the right cost-sharing scheme, larger swaths of the structurally marginalized populations can access the possibility of gutting trailers, stripping them of engineered wood core, adding sufficient insulation, fixing maintenance issues that exacerbate chemical and fungal exposures, and mitigating much of their environmental and climate hazards. The home inspector with the maintenance co-op, whose van I had been rattling around in the back of, admitted to me that there was still much work to be done around the health, safety, and climate readiness of their homes. And hopefully I can join them in that.

Hello

Over the course of this book, you will encounter many people who will share deeply intimate betrayals, stories of grief, dreams, or just the minutiae of life. Under the circumstances, I should tell you at least a little bit about myself.

I was born in St. Louis, Missouri, to a Protestant mother from Baltimore and a Jewish father from New York City, both white. We lived there until I was seven, when my dad got a big break in Los Angeles. That break quickly broke, and we moved back to St. Louis nine months later to regroup. After two years of trying to figure out the next steps, we moved to Atlanta, where I lived until I was eighteen. This is where I think of when I hear the word *home*.

We lived an upper-middle-class life in midtown. My mom taught middle school and my dad worked at and later ran a museum. It might have been moving around a lot that made me more of a quiet observer, trying to figure out how things work, while also self-entertaining to the point that I got lost in my own thoughts. My teenage years were a mix of privilege and exposure. I attended private school, but also had friends and neighbors who died of overdoses. My high school was on the road that served as a key racial fault line; the streets that intersected it changed names so the addresses of white people would never be confused for those of Black people.

I had a sturdy safety net to return to even if weekends were charged with a minor proximity to violence. The high school party circuit in the early 2000s had created social scenes beyond any one school, where house parties would pull in people from across four to five predominantly public high schools. Eventually kids got entrepreneurial and started renting abandoned office parks, stables, and warehouses for massive parties. There was some handgun fire or brandishing, and eventually adult security guards were hired to wand and pat people down at the entrance of our illegal parties. I somehow was at the helm for throwing the big graduation party my senior year. My first-ever text messages were with the friend I was organizing with who ended up in the ICU because he got curb-stomped before the party and couldn't speak.

In the summer of 2005, I was helping some friends from Texas drive their cars to Upstate New York for our sophomore year of college. Oblivious to the incoming hurricane, we stopped in New Orleans for only one night. The next day we exited the city in torrential rain over a pair of low-lying bridges, known as the twin-spans. One of my friends broke down crying, convinced we were going to be blown into the Gulf. Large sections of the twin-spans were later upended by the storm surge on its way to flooding the city. We, through no credit of our own, were unscathed.

Having first seen the city just before it was devastated demanded that I return to it, but I didn't make it back until the spring of 2009. My friend who thought we were going to die on the twin-spans was working in the kitchen of the hotel where we had stayed four years earlier; another one of the friends I was visiting had quit her job in construction and landed an entry-level administrative position at a law firm. Her days happened to be largely focused on the FEMA trailer product liability case. That was the beginning of this project.

My whiteness and the fact that the largest amount of organizing done around formaldehyde in the FEMA trailers took place through white networks in Mississippi skews the people I spoke to in the researching of this book. I undertook the ethnographic work in New Orleans at a time when the city was grappling with a second inundation, that of young educated white people streaming into the city. Many of the poorest Black people never returned after the storm (M. Davis 2010).

Rightly or wrongly, to me the city felt full of white, often northern, people trespassing into Black space in ways that were framed as altruistic but sometimes appeared to border on voyeuristic (Shapiro 2013). As a result, I chose to largely avoid recruiting people to talk to through nonprofit organizations that were full of transplants like myself looking for ways to connect with native, often Black, New Orleanians. My own perceptions of what was racially appropriate also inflected whom I felt comfortable "bothering" repeatedly for an interview. I could have and should have found ways to recruit in a way that didn't bias toward my own white networks because they felt less extractive, exoticizing, or serializing. Attempting to avoid perpetrating, what I viewed, as disrespectful behavior is not an alibi for telling a story that is at least partially racially lopsided.

Behind the Stories (Methods)

One of the fundamental challenges at the outset of this book was how to study a community that pushes the boundary of being a community, as its members have largely never met each other. Most studies of and campaigns for environmental justice have focused on place-based communities over the distributed groups of people that make up communities based on shared consumption that leads to shared exposure. Even this split between place-based and distributed harms doesn't fully hold when thinking of the strange commodities of the mobile home, perched somewhere adjacent to both real estate and automobiles, anchors of senses of place and movable products. The FEMA trailers are serialized places, functionally indistinguishable and often hundreds of miles away from each other like a tract housing development that was cut up and scattered across the country. The traditional place-based methods of studies in environmental justice provided only slight guidance for researching a community of nearly identical spaces that spanned much of the continent, from Puerto Rico to Alaska. The situation even differs quite dramatically from the distributed publics who live alongside infrastructural projects, like pipelines. Despite what I am told is a deeply annoying ability to point out FEMA trailers from the highway on road trips, locating FEMA trailer inhabitants one by one was not a viable method. Instead of finding FEMA trailers, I needed them to find me.

To accomplish this inversion of me as the tracker to the person being tracked down, I collaborated with Kim and Mike Fortun's interdisciplinary research group, known as the Asthma Files (cf. Fortun 2012: 453–56). Kim Fortun provided material support and introduced me to a private indoor air-quality lab—Prism Analytical Technologies, Inc.—that lent in-kind support in the form of instrumentation and chemical analyses.[16] Martin Spartz, PhD, an analytical chemist at Prism; industrial hygienist Linda Kincaid, MPH (featured prominently in chapter 2); anthropologist of science Brandon Costelloe-Kuehn, PhD; and I designed a formaldehyde testing protocol that could be performed by me or mailed to FEMA trailer inhabitants that were too far away to visit. The protocol held constant as many variables as possible (see appendix 1).[17]

The Arkansas-based environmental activist Becky Gillette featured our call for participants on her website (toxictrailers.org) and also placed my email address in the website's top banner. Various news organizations pushed the call through their networks. I lay in wait. At first the emails and phone calls only trickled in, but within three months the rate of response picked up and plateaued at one to three new solicitations every week. Over the course of the next few years, I would be contacted by over two hundred people seeking information about domestic chemical exposure, the vast majority of whom were residing in resold FEMA trailers. Beginning in the spring of 2011 and extending to the spring of 2012, I undertook over a dozen road trips to meet with residents of resold FEMA trailers. I drove several thousands of miles to conduct fifty-one in-person interviews with trailer residents in twelve states and double that number of phone interviews in fifteen additional states. I continued to test trailers until late 2019.[18] I additionally interviewed thirty-five people (approximately half white and half Black) about their experiences living in FEMA trailers after Hurricanes Katrina and Rita, with dozens more brief anecdotes or stories finding me every month during my two years in New Orleans and Mississippi. All interviewees are given pseudonyms by default, but some preferred to use their real names, which I did.

I also attended trailer auctions, visited FEMA trailer staging areas, visited the offices of exposure scientists, participated in and observed domestic formaldehyde testing, argued about chemical risks with FEMA trailer resellers, sat in on a closed meeting of the community advisory panel of a federally funded study to assess the FEMA trailers' effect on children's health, and observed a community meeting where this study was presented to the community. I visited former FEMA trailer inhabitants in their rebuilt homes in Mississippi and Louisiana or in new homes hundreds of miles from the coast—as far afield as Massachusetts—and in hospital beds.

In addition to participant observation and interviews, I consulted thousands of pages of affidavits, deposition transcripts, medical records, and other legal documents as a law clerk in the office of a plaintiff attorney who was working on litigation regarding the FEMA trailers and formaldehyde product liability. I also wrote half a dozen Freedom of Information Act (FOIA) letters to federal agencies requesting data on both the original deployment and the resale of the FEMA trailers. One took over six years for me to receive the data I requested. I made a mini-documentary with Mariel Carr about the resold FEMA trailers on the Bakken shale field and nearby Native Nations that has been viewed more than 4.7 million times (Carr 2015). I worked with Grist, an environmental news nonprofit, to make a vehicle identification number (VIN) lookup tool with documents from my FOIA requests that enabled people who wondered whether their homes were FEMA trailers to check the origin. People still contact me about these issues a decade and a half after I began this research, albeit at the reduced rate of once every few months. Some projects that inform how I theorize change in this book and that were nearly all-consuming for years barely surface at all (Shapiro and Gehrke 2015; Dillon et al. 2018, 2019), as the point is not to tell a story of what I've done, but to communicate my observations of what kind of work can produce what kind of impact on the various homesicknesses of the United States.

I came to these methods by taking seriously Kim Fortun's (2001: 6) charge to engage with multiple strata of data. In science and technology studies, the collaborative methods I outline above are often known as "making and doing." I find great inspiration from my colleagues in this field (to name just a few: Choy et al. 2009; Vera et al. 2020; CLEAR 2021; Downey and Zuiderent-Jerak 2021; Liboiron et al. 2021; Schroeder et al. 2021) even if the moniker doesn't sit quite right with me. Writing, making scientific instruments, narrating documentaries, and building remediation tools are all simply different ways of thinking about an issue. I didn't understand this until Tim Choy (2011) pointed this out to me years ago. I write until I hit a limit, engineer until I hit a limit, write again, limit, remediate, limit, art, limit, and so on. Each mode of inquiry helped broaden the standpoints I can represent in this book. It has also helped me feel more confident that the countless hours that exposed people spent talking to me was worth their while.

FOR HALF A CENTURY, the potential crisis of indoor formaldehyde, largely emanating from the engineered woods it holds together, has emerged with an episodic frequency. European and Middle Eastern chemists who wrote "Formaldehyde in the Indoor Environment" in the journal *Chemical Reviews* noted that "every few years, the 'formaldehyde discussion' [in the United States] is resuscitated" (Salthammer et al. 2010: 2565). *Every few years*, that is, systematic domestic chemical exposures break into the discursive foreground only to be quelled by perfunctory reviews, regressive policy, or, more often than not, the simple ebb of time. *Every few years* formaldehyde fails to crest cultural horizons of significance despite the FEMA trailers' service in the wake of the costliest natural disaster in US history.

Beyond the silent hazards of chemicals, more than thirty million US households, over one in five, live in the presence of significant physical or health hazards, such as dilapidated structures, poor heating, damaged plumbing, or gas leaks. Only 3 percent of public housing units that are home to a disabled member actually have a unit with accessibility features (Ross et al. 2016). *Homesick* offers one of many inroads to this larger condition of living in what is now known as the United States, valences of which this nation also inflicts on the rest of the planet.

More than a tale of a chemical and its favorite form of housing, this book is a story about struggles to survive and extinguish the many forces of domestic immiseration. It is a methodological story about how to frame a problem and what kinds of possibilities for change emerge from different understandings of what the problem is. It's about how homesickness was a founding principle and remains a foundational texture of life in this nation, and how homesickness for dramatically different ways of dwelling and relating bubbles up from this surreal misalignment of dream and actuality. It is also an unflinching blow-by-blow history of the state violence advanced by federal emergency housing. Being homesick for an otherwise future helps us think past an environmentalism that regulates chemicals one at a time, past illusory accountability systems, and toward an understanding of the multiple and disparate approaches needed for meaningful change. If living differently is one of the basic challenges of the twenty-first century, where we spend most of our time, at home, seems like a good place to start.

ONE

At Home in the Surreal

In Alabama, I visited a man whose used-car lot overflowed with white boxy Federal Emergency Management Agency (FEMA) trailers. It was March 2011, just outside Birmingham's suburbs. The previous week, several days of intermittent tornadoes had destroyed thousands of homes across Alabama, Arkansas, Georgia, Mississippi, Tennessee, and Virginia in what was arguably the largest tornado outbreak ever recorded (NOAA 2012). Alabama bore the destructive brunt of the storms. Just blocks away from my friend's house in Tuscaloosa, where I was based for the trip, houses were cleanly cut in two. Books lined the shelves of rooms that opened out, like tragic dollhouses, onto empty concrete slabs. Steel transmission towers doubled over like wilted flowers. Whole apartment complexes were disassembled, scattered, and laid flat.

A white rural couple with whom I spoke had barely enough warning from the tornado sirens to jump into their bathtub. The winds hoisted up their home in its entirety, only to release it, moments later, sending it crashing down ten feet askew of its foundation. After the storm, they placed a privately acquired FEMA travel trailer in their front yard and contemplated whether to rebuild or simply downsize to a mobile home.

This same weekend, I found a FEMA trailer reseller's ads online. Together with the investigative journalist from New Orleans I was working with, I set up a meeting with the man, whom I'll call Sid. Sid recognized us upon arrival through the large showroom windows as I sorted through consent forms, clip-boards, and notepads on the trunk of the car. He came out to greet us in the dirt parking lot. A slender man, he wore a T-shirt and long jean shorts with patches of red embroidery running down the outer seams. Unlike the vast majority of trailer resellers I spoke with, Sid was Black and he was candid. He ushered us inside.

Seated at a table in the largely vacant showroom, Sid downplayed the idea that people planned to live in his trailers, suggesting they were primarily being bought for recreational purposes—as lake houses or hunting camps—or simply for storage. As we spoke, an older white man shakily walked into the showroom, his wide eyes magnified by thick glasses to the point that he appeared almost cartoonishly shell-shocked. "I need a home," he declared. He stood by the door, his mouth a small tight line, wearing black moccasins and aged but meticulously maintained jeans. Sid hesitated momentarily. To clarify any ambiguity about his situation, the older man named the tornado-torn backwoods area that he called home. Sid met our knowing looks with a small bow of his head before rushing off to show his new customer the available models.

Sid was no doubt selling FEMA trailers as once-again emergency housing despite their legal designation as unfit for human habitation. When Sid returned, he reasoned with us, "You can put [warning] stickers on 'em all you want, but people are still going to live in 'em if they need to live in 'em." He continued with a solemn look on his face:

> I am just being real, cutting to the chase. The biggest thing is safety, health safety. If they [the government] gave the truth about the health safety, from head to toe, and let the general public know what to look for and what not to look for without going through a whole lotta red tape, then it's all good. But it ain't about that, it's about liabilities here and *people can't do no better*, you got people homeless, sleep under doggone bridges. Where you think they gonna stay? In a trailer rather than under the bridge? [Yeah] they gonna stay in a camper [a synonym for this type of trailer].

In Sid's view, profound rural and ex-urban poverty (as he was selling these trailers before the tornados) drives local housing desperation and the murky, government-generated information that makes it nearly impossible for those desperate people to realize healthy housing choices. Within these conditions created by a government more interested in reducing liability than stewarding health, it was nearly inevitable that some people would use FEMA trailers as homes. Sid maintained that the trailers were built to be lived in for extended periods of time: "The government also knows that people are going to live in them, I mean that's what they gave 'em for. I mean, people stayed in 'em, I mean. The government made 'em for folks to live in. I mean, it's not for permanent housing, you follow what I'm saying, but people *lived* in 'em. Some people lived in 'em more than a year. Ya, ya, ya know?" Every step of the way, Sid had to batten down his meaning. "I mean" peppered Sid's speech when he attempted to distinguish between the trailers' use and their original purpose. He similarly struggled to describe the slippery timelines of inhabitation that fall into the gap between temporary and permanent. Although he laid out his explanation in a straightforward manner, the vagaries emergent in the space between the trailer's function and averred usage made it highly possible that I wasn't following what he was saying.

But just as Sid "cut to the chase," his straight talk, secured by a cascade of refinements in meaning, began to recede. His lot, perched on a hill overlooking a two-lane highway, was adorned with a banner displaying the FEMA insignia and a hand-drawn sign labeling it a "FEMA trailer liquidation site." The ontology of the trailers was woven together not just by architecture and the reseller's description but also by an improvised symbolic association with federal emergency response. Because other resellers uniformly excluded the name "FEMA" from any advertising, I asked him, "What was it about FEMAness that was potentially appealing?" He leaned back in his chair, crossing his arms. "Now, that's a good question. . . . Maybe it is that the people are thinking that they are getting a good deal, which they are, they are saving money. . . . FEMA right now is a pretty good thing for these [tornado-displaced] folks. . . . They see 'FEMA' and it helps us, so be it."

Sid contends that his honest selling cuts through illusory warnings that attempt to re-specify the trailers as something other than what they are. Yet his marketing tactics, particularly following a natural disaster, suggest that his trailers are a sanctioned refuge for the displaced. He benefits from a spurious affinity with the federal government, just as FEMA's first disaster assistance checks were arriving in the tornadoes' wake.

As I tracked the resale of FEMA trailers across the country, these simple homes, composed of the most basic forms with the most basic amenities, began to appear unspeakably complex. Permanent temporary housing. A private place to lay one's head that grips minds with migraines. Housing that is most honestly sold for long-term inhabitation even if the housing's warranty is voided if lived in for more than two weeks straight. A shelter that invisibly suffocates a four-year-old boy on a rural reservation at night to the point that he has to

FIGURE 1.1. Resold FEMA trailer next to an empty foundation and trees stripped of their leaves in Phil Campbell, Alabama, March 2011. Photo by Nicholas Shapiro.

be medevaced by a helicopter from his own home. Emergency housing units that are speciously sold as emergency housing units. Warning stickers that seem to undermine the very nature of the structure. From their illusorily low price tag—subsidized by their often-hidden health costs—to loop-the-loop explanations of why formaldehyde disclosure forms are always necessary yet the chemical is never a problem, this chapter details the viscous surreality emergent in life within toxic homes. Rather than proceeding chronologically, the story begins within the smothering and marshy reality of homesickness. Like much of late industrialism, these exposures corrode rather than directly kill and this fog is almost invariably the starting condition for all the subsequent actions detailed in this book.

Failures of matter, meaning, the market, and the state are entangled in the production of this domestic surreality. The surreal character of homesick life emerges from the mysterious chemistry of low-cost building materials that left an Alabama woman wondering if the vinyl siding of her trailer would melt off, or a Texas woman continuously running her hands across the small blisterlike protrusions sprouting from her walls, trying to smooth them out. From sunken-in floors in Oklahoma. From shoddy wiring in Mississippi. From manufactured homes that were assembled by joining two identical halves (as opposed to two corresponding halves), producing a confusing architecture of ill-formed spaces, accidental floor plans, and anti-ergonomic frustrations. From chemical exposures emanating from the walls of the home and the airs of suspicion that such invisible toxicants ignite. From the material and imaginative decay, a defamiliarization of one's place in the world.

MORE THAN JUST A catch-all for the residual out-of-place moments we might casually refer to as "surreal," in this chapter I aim to show how an analytic of the surreal can help make meaning from the altered states, illogics, and material breakdowns that emerge from ordinary and extraordinary encounters with changing environments and extractive economies. In these accumulations of incoherent quasi events and the failure of the home's, the state's, and the market's symbolic protection, reality bows into the surreal (cf. Cowie 2007: 206). A toxic home itself is not a surreal thing; rather, it furnishes a surreal state of things, an everyday in which meanings, matter, and knowledge are left ajar, a usual unusualness that is marked by material, biological, and semiotic decomposition. Surrealism is an appropriated genre of reality that emerged from the experiential friction between official proclamations of triumph after World War I and individual trauma of those freshly returned from the death-worlds of the trenches (Clifford 1981: 541).

But it is also a key texture of racial capitalism, as markets create shocking juxtapositions and rearrangements of the world's things. As Robin Kelley (2002: 165, 169) has incisively noted, the "absurdity of racism and the fragile and strange world of white supremacy" made surrealism more "confirmation than a revelation" for Black writers like Aimé Césaire and Richard Wright. The relentless illogic and cleaved world of racism serves as a bridge between the figures of the state and market as racial capitalism supplied these trailers primarily to those excluded from whiteness.

A far cry from the giddy artistic leveraging of "hallucinatory forces" to "breach the order of mechanistic processes" (Einstein 1929, in Clifford 1981: 549) that delighted European surrealists, the surreality of North American homesickness is charged by short circuits, deterioration, settler colonial structures of land tenure, and chemical off-gassing—a reality that wavers between tedious, bizarre, and nightmarish. Nightmares were themselves one of the most common symptoms associated with living in resold FEMA trailers. Increases in nightmare frequency and intensity were experienced by one in four of those with whom I spoke.

This chapter moves from the bewildering warnings and illogical policies surrounding the initial deployment of the FEMA trailers, warnings and policies that residents themselves identified as surreal, to technoscientific assessments that are distinctly absurdist. Instead of providing a path for mending hazardous living conditions, quantitative evaluations often manipulated conditions, produced dissonance, and rubber-stamped hazardous homes as safe. The dangers of these homes are not limited to chemical exposures, but include a whole host of hazards produced in tandem during the rapid and flimsy manufacture of low-cost housing. Electrical short-circuiting, fire risks, warped structural elements, and toxic air together condition the sustained surreality of these homes, as both indicators and portents of the larger infrastructural and climate instabilities of a world built by racial capitalism.

The Treachery of Warnings

The range of amenities a house is thought to provide (safety, security, and health benefits) do not always accompany the visual appearance of a home (Lea and Pholeros 2010). And, indeed, notices warning not to confuse these housing units with homes began appearing on the FEMA trailers during their initial deployment. I met Mominem, a white former FEMA trailer resident, in the meeting room of his New Orleans–based architecture firm in June 2011. He recounted the bewilderment of receiving his first warning of potential bodily harm resulting from his FEMA trailer: "I'd been living in the trailer for two years and they [FEMA's contractors] came along [while I was out] putting new warning stickers on in three languages: English, Spanish, and Vietnamese." We both laughed.

In some ways laughter is the only logical response. The advisory notice was comically thorough in its inadequacy, using three languages to apprise trailer inhabitants about harmful chemical exposures over a year after FEMA unequivocally knew their emergency housing units were hazardous. Most inhabitants were already moving out. The temporal raison d'être of warnings, their ability to avert impending danger, had progressively decreased in strength while human bodies soaked up formaldehyde for over two years. All that was left was the state's vacant language of liability mitigation, available to salt wounds in multiple tongues.

Living alone in his FEMA trailer while his wife remained evacuated in Atlanta, Mominem turned to blogging to pass his nights and process the strange happenings of post-Katrina New Orleans. His blog maintained a thread of playfulness despite the stress of living in a battered and broken city. He chalked up his lighthearted air to "the absurdist situations that we're in all the time." Mominem believed the bizarreness of governmental interventions and their resulting bafflement was best received with a laugh.

In his blog, Mominem voices his bewilderment following FEMA's disuse of over ten thousand mobile homes during postdisaster reconstruction efforts in 2006.[1] These mobile homes were stored at the Municipal Airport in Hope, Arkansas. The stock of homes outgrew the stable ground of unused tarmacs and taxiways, so that three-quarters of the units were stored on adjoining soybean fields (Neuman 2006). When it rained, the homes began to sink into the mud. Some bowed. Other units began to topple into each other, as they were parked as little as six inches apart. FEMA then spent $4.2 million to cover the field in gravel to stabilize the units. During this same time, 60 percent of Louisiana's trailer requests remained unfulfilled, and one hundred Mississippi families remained in military-style tents (Committee on Homeland Security and Governmental Affairs 2006: 6, 27).

Some twenty-five years before the mass purchase of these mobile homes, FEMA had adopted a protocol that forbade the placement of mobile homes in areas designated as floodplains. The bureaucratic logic behind this policy was that mobile homes could not be evacuated in the event of another hurricane, destroying the agency's emergency infrastructure. But because virtually all of alluvial southern Louisiana is a floodplain, and FEMA refused to bend or change its rules, the units sat, uninhabited, in muddy bean fields. FEMA deployed smaller and more readily portable travel trailers in their place. Ironically, the travel trailers were installed, all across the Gulf Coast, in ways that eliminated their supposed key virtue, their mobility. The travel trailers were jacked up on cinder blocks, tied down with metal straps, and connected to the local electrical and sewer systems. The thought of disengaging 100,000 travel trailers from their sedentary perches and evacuating them in the three or so days of hurricane lead time, simultaneous to the rest of the regional withdrawal, is a telling farce. As Mominem wrote in his blog in 2006, "FEMA as usual has come up with a completely dada series of rules regarding the placing and removal of trailers."

One way of understanding the genesis of absurdist policies like this, beyond raw incompetence, is to recognize that these and similar policies may be driven more by affect than effect. In her ethnography of Australian bureaucrats, anthropologist Tess Lea asserts that "senior policy makers thrive on the emotional thrill of surfing crises," in a way that obscures the impracticality of recommended measures (2012: 111; 2008). This insight leads her to claim that

FIGURE I.2. The negative space of a removed warning sticker on a former FEMA trailer. A finger smudge from removing the sticker is visible. Photo by Nicholas Shapiro.

the state is, counterintuitively, a seat of anarchy. These anarchic qualities are transmitted, not countered, by way of crisis intervention.

If one applies Lea's insights on Australian bureaucrats to governance in the United States, it becomes less surprising that the warnings affixed to the FEMA trailers by the General Services Administration (GSA) prior to their resale appeared to invite disregard. The notice of hazard on the FEMA trailers is not riveted into the siding or etched onto the tow hitch next to the various other identifying marks embossed there. Rather, the warning takes the form of an 11.5- by 3.75-inch sticker on one of the windows that states "NOT TO BE USED AS HOUSING." Glass is one of the trailer surfaces from which stickers of any kind can most easily and cleanly be removed with a sharp blade and a bit of spit. The less-detail-oriented resellers simply rip them off, leaving behind a patch of grime-free glass outlined by a white ribbon of adhesive (fig. 1.2). The rapid removal of these warnings, ripped off like Band-Aids, warps the authoritative performance of their application. Of the approximately one thousand resold FEMA trailers that I encountered in the course of my research, fewer than fifty bore warning stickers.

The red-on-white stickers adhered to the FEMA trailers dissimulate their function. The warnings dissimulate that the trailers were designed to house families of four, or sometimes more, for extended periods. NOT TO BE USED AS HOUSING. The stickers attempt to, at least ephemerally, separate the immediate functional value of the trailers from their use. The warning stickers were applied exclusively to a single window of each FEMA travel trailer and park model. They were not applied to mobile homes.

These warnings are far from being a regulatory outlier. The contemporary landscape is strewn with minute regulations, like the Prop 65 cancerogenic substance warnings posted on every outdoor food vendor in Disneyland, on children's toys, and across entire housing subdivisions.[2] As critical legal scholars have noted, we are exposed to such an "array of injunctions, warnings, directions, and threats that the breaking or avoidance of regulation becomes expected and normal" (Hermer and Hunt 1996: 457). The superlatively shambolic nature of the FEMA trailer warning sticker is revealed in its impermanence. The stickers attempt to garner a permanent disclaimer of federal liability while only temporarily discouraging their use as housing.

At the GSA auctions, more warnings surfaced at the contract-signing table. Purchasers of "usable" FEMA travel trailers and park models were required to sign sales certifications stating that the buyer is aware that the unit(s) they are purchasing "may contain formaldehyde" and that they agree to "hold the United States harmless and indemnify the United States from any or all debts, liabilities, judgments, costs, demands, suits, actions, or claims of any nature that may arise from or incident to the sale of this property its use or its final disposition." While these documents are warnings in that they indicate the systematic harm that may result from selling these units, their cautionary element appears to be secondary to their liability-reducing function.

Virtually all trailers and mobile homes come with some notice of potential formaldehyde presence, as such housing has historically borne, on average, four times as much indoor formaldehyde as conventional homes (COEHHA 2001). The GSA warnings do not distinguish themselves from these all-too-common warnings. They do not indicate, for example, that the FEMA trailers were found to contain formaldehyde levels two to five times higher than the already-elevated average manufactured home (CDC 2010). The notice dissolves the specific risks of living in a FEMA trailer within the systemic and ongoing risk of dwelling in a manufactured home. The notices further black-box potential harm by lumping it with the unlikely possibility that the trailers may not contain any formaldehyde at all—a point emphasized by resellers and contradicted by the chemical samples I collected.

The sale certificates for the FEMA trailers included a mandate that subsequent buyers be informed that such units were not to be used as housing. However, this provision is not enforced. The GSA's Office of the Inspector General either ignored or dismissed those FEMA trailer dwellers who filed complaints. One white Tennessee woman who was renting a former FEMA unit in a trailer park as her retirement plan was lucky enough to receive a response letter from the GSA. She summarized it for me over the phone: "This problem doesn't really have anything to do with our department." Beyond the pattern of chaos, beyond Dadaist interventions, beyond the contested ontology of the FEMA trailer is a stolid, attorney-reviewed, bases-covered disavowal of responsibility. Material dysfunction and biotic decomposition followed the US government's calculated exit from liability.

Rip Out the Walls

Dorothy Keme lives with her husband, Franklin, two of their grown children, one son-in-law, and six grandchildren in a former FEMA trailer on a reservation in Washington State. Since moving into their mobile home after their conventional home burned down in 2011, the family has been wading through simmering headaches. Bloody noses became such an established part of life for two granddaughters that they became unremarkable, like yawns. One grandson was regularly sent home from preschool, as he would split his time between the classroom and bathroom because of diarrhea, a symptom shared by many in the household. "I was concerned when the issue of the formaldehyde in the trailers came out, but you know what? We have nowhere else to live," Dorothy relayed over the phone in almost a whisper. She continued, "My husband has some plan to rip the walls out . . ." trailing off into a heavy silence. The destruction of their home is the only way they could imagine it as habitable.

Delvin Cree, an enrolled member of the Turtle Mountain Band of Chippewa, had been trying to prevent situations like this since 2008. Cree traveled down to Purvis, Mississippi, in the summer of that year to haul a donated FEMA travel trailer up to his territory in North Dakota with his father and brother-in-law. Within minutes of entering the unit, his eyes and throat began to burn, and misgivings about these "gifts" began to percolate. Thousands of dollars in gas and accommodations later, they had towed the trailer up to its new home. Once they got there, water seeped into the trailer's ragged seams when they washed down the outside. It wasn't long before Cree discovered mold in the walls and under seemingly new flooring.

Cree sent photos of the mold to the Senate Indian Affairs Committee and sounded the alarm via op-eds in Indigenous media (e.g., Cree 2012). Plans to house homeless tribal members on his reservation in these travel trailers fizzled. Due to formaldehyde's ability to affect glucose metabolism (Tan et al. 2018), Cree additionally worried about how these gifted shelters would impact the many diabetic tribal members slated to inhabit the homes. The roughneck gentrification of the nearby Bakken oil and gas fields had kindled an acute housing shortage at the time. Those flocking to the state to build up a nest egg or pay off debt with black gold, in a strange twist of fate, ended up renting or buying many of the toxic trailers donated to nearby Native Nations.

Standing by a dilapidated FEMA travel trailer on his relatives' property in a striped polo shirt, Cree recounted previous ill-fated free tribal housing schemes, such as refuse military housing that contained asbestos. He viewed the massive distribution of trailers not as aid from benevolent occupiers but as a dressed-up means of disposing of the trailers' liabilities: "I think that's why the government gave us these camper trailers, because we can dispose of them on [our] own lands."

Yet, many Indigenous recipients of these trailers, like Dorothy, were plunged into exactly the kind of domestic medical chaos that Cree had worried about. They lived in not only the smaller camper trailers, but also the larger FEMA mobile homes that federal officials assured were safe. Charline, a cook for Head Start (a federal program for early childhood education, health, and nutrition for low-income children) on a reservation in South Dakota, first wrote to me in late October 2012:

> I purchased a FEMA trailer in 2009 and I've noticed that ever since I moved in here my children have been always sick. My son who was little over one years old and is now four years old has been in and out of the hospital for breathing problems, has occasional nose bleeds, and has just fallen over at school twice. My daughter who is two years old now has been hospitalized numerous times with breathing problems and was even ambulanced out before because of respiratory problems. Now my son who is seven months old is having the same breathing problems. He was recently flown out because they thought he had pneumonia, his oxygen levels were very low. Then after spending one night in intensive care and four nights in the peds unit he came home and then about two weeks later he was hospitalized again because of his breathing. Myself, I have noticed that I have had terrible migraines nosebleeds and repetitive sinus infections.

I tried to track her down when I was in the Dakotas, but the phone number I had for her had been disconnected. Her old employers didn't know how to reach her. In early 2021 Charline wrote to me to ask how to buy a copy of my book and let me know that she and her children were able to move out of the trailer in 2016. Shortly thereafter, all of her children's distressing health problems evaporated.

Safety by the Numbers

After Hurricane Katrina flattened Brooke and Ronnie Granger's home, FEMA provided the Picayune, Mississippi, residents with a mobile home—the more spacious type of emergency housing unit that was disallowed on Louisiana's flood-plains. Brooke, fifty-three, and Ronnie, sixty-three, both white, were living off So-cial Security and retirement benefits and couldn't afford to move into a more per-manent dwelling. In 2008, FEMA offered to sell them their emergency housing unit for $5, a standard price for the first wave of resold FEMA trailers. At that time, the agency calculated that taking the trailers back and reselling or storing them for future use would likely cost more than their replacement value. Despite experiencing a slew of health issues that had emerged or accelerated since they moved into the trailer, the Grangers jumped at the opportunity to purchase what was already their home (fig. 1.3). The proposed exchange came with the stipulation that the mobile home would be tested for formaldehyde and must register low enough to be deemed "safe" for the transaction to be finalized.

Records kept by the Grangers show that FEMA's first formaldehyde test registered a result of 77 parts per billion (ppb), almost ten times above the level that the Agency for Toxic Substances and Disease Registry states ushers in detectable health risks when inhaled chronically (see appendix 1 for more exposure standards). Closer to the winter sale date, FEMA came back and asked that all the windows and doors be opened for a retest. "We're going to let it cool down. We want it aired out," Ronnie was told over the phone. Brooke and Ronnie sat quietly in coats and gloves on the couch as they complied with the new test. Formaldehyde off-gassing is tightly correlated with temperature and can be heavily diluted with uncharacteristic ventilation. Accordingly, the sec-ond test result was 17 ppb, low enough for the sale to proceed. In August 2010, a third party tested their home with a twenty-four-hour badge and found a formaldehyde concentration of 117 ppb. In October 2011, my own one-hour formaldehyde test of their home produced a result of 105.69 ppb.

Since moving into the trailer in 2006, Brooke and Ronnie have experienced a slew of health problems associated with chronic formaldehyde exposure: crumbling teeth; changes to their sense of taste; distortions of their voices, eyes, and respiratory tracts; skin irritation; and chronic obstructive pulmonary dis-ease (COPD), among others. Three of their dogs have died of kidney, liver, or bladder cancer. Two pet birds also died abruptly. Their vet states with conviction that formaldehyde caused the death of all five pets, yet admits he can't prove it.

I visited the Grangers three times. Each visit lasted for a couple of hours, and every time we sat on their bed, which was placed in the living room. The

FIGURE 1.3. The Grangers, near Picayune, Mississippi, December 2011.
Photo by Andy Cook.

power in the master bedroom had long since ceased to function, as had the power in the bathroom and second bedroom. A master electrician said the whole house needed to be rewired; as a stopgap remedy, he ran extension cords underneath the house. The electricity in the living room comes and goes. Two outlets were replaced after they spontaneously burst into flames. One of the plugs that caught fire was situated below where they hung their family photos. The photos have since been moved to plastic tubs by the door. Many of their valuables sit in those tubs, Brooke explained to me. Some things she never unpacked, as she didn't want to be in the mobile home long, even though they had been there a decade when I last spoke with them. She wanted to be able to quickly save other items in the event of a fire. She worried about fire constantly. Ronnie did too: "We live in fear of fire every day."

Brooke's dental health has deteriorated. "My teeth have been getting bad, just crumbling out of my mouth. This one [pointing to a lower left premolar] is gradually crumbling and breaking off. It's crumbling from the inside." Sitting cross-legged on their bed, she recounts, "[My dentists] can't say 'yes' and they can't say 'no' but they say that the formaldehyde is a very good possibility of a cause . . . and the way that it is happening, it's not just rotting and falling out, it's good teeth."

While chemical assessments representing the concentration of a known human carcinogen in their domestic air fluctuated by a factor of 6.9, Brooke's and Ronnie's symptoms remained resolutely consistent. Numbers, warped by the prerogatives of federal liability, belied the regularity and depths of their domestic harm. These exposures, and the government's disavowal of them, left them bracing against their shelter and continually expecting future shocks to emerge from a world beyond visibility and certainty.

Stable Ground

Two journalists accompanied me on a second visit to their home. The journalists were developing a story about the Grangers' situation. On our way back to New Orleans, we stopped by a sprawling field of almost indistinguishable, unoccupied FEMA trailers. Andy, the photojournalist, wanted a photo of the scene from an elevated viewpoint. I scurried up the ladder on the back of a trailer to take the shot for him. As I moved from the ladder onto the roof, the trailer's corner gave way. The waterlogged and rotted plywood was as supportive as dough. There was no squeal of wood. No pop. I just sank in until there was nothing left beneath me. A puff of mold spores exited as I fell in. I had never been inside that particular model, but I had been in enough FEMA trailers to know that if I fell to my right, I would land on the top children's bunk. If I fell to the left, I would tumble into the shower. Leaning right, I only fell up to my knee. My ephemeral self-satisfaction with navigating the fall shifted into fear as I scurried down from the roof of a trailer about to decay into itself. The interior was coated in a thin fungal fuzz as the unit appeared to have long been flooded.

In suburban Texas, I tumbled out the door of another trailer—this one occupied—and down the two-foot drop to the concrete when I assumed that the floor under the welcome mat was intact. It had rotted out. My teeth and, to some extent, my dignity, were saved by landing on a cooler that the trailer owner had placed just outside the door and was using as a sink. The shoddy plastic tubing that constitutes the trailer's plumbing had blown. For the past few months the white woman who called this trailer home had been washing her dishes in the cooler outside, spritzing them with a garden hose (fig. 1.4). She rushed to the door, looking down on me as I lay sprawled over the cooler. With urgency, she asked, "Are you okay?" I responded in the affirmative. "It gets me all the time," she remarked as she retreated back inside to where we had

FIGURE 1.4. Washing dishes in Pasadena, Texas, November 2011. Photo by Nicholas Shapiro.

been watching *To Kill a Mockingbird* on TV. Some pitfalls are so regular, so unavoidable, they are not even worth a warning.

The hazards of material disintegration and the undercutting of confidence that they generate are shock-absorbed by inhabitants who don't have the resources to resist them. For work, this woman in her early fifties assisted the white, pain-pill-addicted, elderly widow who owned and ran their trailer park. Her only income was the $100 a week paid to her by the park owner and the in-kind payment of free water (but not electricity) and a free place to park her trailer. She had no phone or car. Her freshly divorced nephew, a thirty-seven-year-old welder, had recently moved in with her and slept on a pallet on the floor. I came to know this woman after I recognized her trailer as a FEMA unit, having been contacted by another FEMA trailer resident in her mobile-home park. It is unlikely that she would have found the time and access to the internet necessary to track me down on her own. She had largely given into the ordinary surreality of the trailer. I was alone in feeling a sustained sense of the surreal, while she busily made do. A formaldehyde test I conducted in her trailer yielded a result (105.69 ppb) significantly higher than the World Health Organization (WHO) recommendation for formaldehyde exposure limit for thirty minutes (81 ppb).

Markets

In the spring of 2011, I visited a trailer park twenty minutes outside Houston's perimeter highway. This park had acquired just fewer than one hundred FEMA trailers, a quarter of which were designated as scrap.

Dirt mixed with green mildew edged up the sides of the trailers on this perhaps five-acre business. The park consisted of two lots. One hosted the office, a remodeled modular home fitted with a generous deck out front, and trailers to rent on-site. The other lot was dedicated to reselling used mobile homes and trailers. In the faux wood–paneled office of the owners—a white husband-and-wife team—we rehearsed the usual heated exchange that I was accustomed to with resellers. I would press them on why formaldehyde is always an issue, always warranting a legal release but is not, in their view, a cause of illness. It would usually go nowhere. That visit was no exception.

As the husband, Daryl, walked me out of the office, I asked him how he got started in the trailer business. With a prideful smirk and a mysterious comment, "Let me show you something," he led me to the front office and a small collection of black-and-white photographs from the 1950s, featuring a high-end mobile home (fig. 1.5). "My grandmother was Mrs. Mobile Home in Florida in 1952," he proudly said, gesturing to a yellowing newspaper clipping. "My grandparents designed the first mobile home in Florida, 'The Futuro.' It was all custom. Here ya got: them, cousins, and the investors. Mobile homes are in my blood." Flush with pride, he recounted the particulars of each shot, the model's features, and every one of his depicted relations. After a brief but wistful pause, he rushed out to whisk his son off to his baseball game.

I stood in their front office, staring at the collection of images. Beaming men proudly standing by the first Futuro to come off the line. Airy shots of the mobile home's interior: ample seating; some built-in fixtures for maximum efficiency, convenience, and sturdiness; freely movable stools, chairs, and end tables to lend the feeling of a conventional home. An illustration showed the Futuro hovering in a white abstract landscape, replete with design flourishes similar to the more expensive automobiles of the day: A series of gill-like windows encircle the home, lending it some resemblance to a spacecraft. And finally, the centerpiece of the constellation—the Futuro in situ on a secluded beach, palm fronds leaning into the frame from the right.

As I stood in the reception area, staring at nothing less than a shrine to the Futuro, I got lost in the expanse between the sixty-year-old dreamworld of mass-produced comfort and escape and the seemingly inescapable hardscrabble realities of the current mobile-home business that surrounded me.

FIGURE 1.5. Photographs of the Futuro, on a wall in a trailer park outside of Houston, Texas. Photo by Nicholas Shapiro.

The world of the Futuro appeared nearly antithetical to the business that Daryl had been describing to me minutes earlier, a description full of backpedaling, grimaces, sighs, and running his hand down his cheek to his chin slow enough that the rasp of each follicle of stubble as it passed under his hands was almost audibly distinct. Times were no doubt hard in the FEMA trailer resell business.

A comment from the white receptionist punctured my thoughts. "Ya know, those Louisiana people bought Cadillacs with their FEMA money," she called out from across the room, with more than a hint of venom. The phantasmagoria of the Futuro lurched to a halt. I returned from the nostalgic futurism of the photos to my current location, a shag carpet that resembled a dog's coat after a day at the beach. To the mildewing shelters of last resort that the proprietors rented to the down-on-their-luck, the newly evicted, or the otherwise structurally marginalized. To a manufactured interior space where nothing is custom-built. To the news story about Daryl buying a mobile home with a dead body in it, which hung discreetly in a frame in the corner. To the grounded dream of the mobile home.

In Texas, especially, white people use the term "Louisiana people" as a dog whistle to refer to Black people without saying more obviously racist words. The receptionist invoked the term as cover for the racist trope of emergency-relief welfare queens; of course, this stereotype is the inverse of the truth. Across the United States, the more FEMA money a county receives, the more white wealth grows and the more Black wealth declines (Howell and Elliott 2018). She would not be deterred by this reality. "Cadillacs got formaldehyde in 'em too, why couldn't that be what's making 'em sick?" she continued, pursing her lips and raising her eyebrows when I turned to look at her. "Well," I thought out loud, "you don't live in your car—I mean, you don't spend as much time in it.... It's easier to ventilate... um, there is probably less formaldehyde in it because there are no engineered woods in 'em... and how many people really bought Caddys, especially new Caddys, with their FEMA money?" She waggled her head disapprovingly at my answer and turned to her paperwork. A gust of asphalt-singed air blew into the room as a Latinx woman pushed open the front door, a baby slumbering on her chest. "Do you have anything I can move into today?" she asked in a thick accent as I left, pushed out by the receptionist's glare.

As I thought more about it, the jump-cut from the immaculate whiteness of a past future that did not include people of color—or the poorest whites—to the gritty racist and classist present, resolved to a smooth transition. From exclusion to predatory inclusion, the dream of the Futuro and the reality of the FEMA trailer occupy different instantiations of the same racial capitalism. The loop-the-loop reasoning of waivers for allegedly nonharmful exposures and the illogics of racist attempts to dismiss previous (or simply imagined) Black residents' exposure in the name of underlining their safety today all texture the surreality of low-income housing under racial capitalism.

In the context of FEMA trailers, this outpost represented mid- to small-scale reselling, yet still cost the buyer over $215,000. FEMA had begun trying to offload some of its Katrina-era trailers as early as 2007, but the market, at that point, remained small. For example, documents I acquired through public records laws indicate that a white man named Wade Haldane bought seventy-five FEMA travel trailers from federal auctions in February 2007. Wade set up a limited liability company outside of Tampa, Florida, that he named "E-Z RV" and began selling his trailers on eBay. He used deceptive photos and descriptions. One man, who specifically asked Wade if the trailer he was looking at online was a FEMA trailer, was told that it wasn't. When the man demanded a refund after spending over $400 on fuel to pick up what absolutely turned out to be a FEMA trailer, Wade agreed—but only on the condition that the buyer didn't leave negative feedback on eBay.

Another buyer, Meredith, a retired schoolteacher in Missouri who identifies as both white and Indigenous, unknowingly bought a former FEMA trailer from Wade in June 2007 for $6,035. She thought it was a bargain.[3] When news of the formaldehyde issues broke a week or so after she bought it, she immediately had her trailer tested for formaldehyde. The test revealed an indoor concentration of 254 ppb, more than triple the WHO's maximum exposure guideline for half an hour of exposure. Meredith pleaded with Wade for a refund or explanation over eBay's messaging system. Wade not only ignored her pleas, but also quickly shuttered his company to avoid litigation. The trailer still sits on Meredith's lawn, growing mold and slowly sinking into her yard. She calls it her "shameful lawn ornament."

In July 2007, FEMA launched a short-lived federal refund program built on a surreal vision of restitution for people like Meredith, who bought their trailers before the agency acknowledged the formaldehyde issue. Importantly, it was a refund program, not a recall, meaning that the government would compensate trailer owners who surrendered their trailers at the site of purchase, but it would not cover costs for transport, which sometimes equaled or exceeded the initial price of the unit. Even if we play macroeconomist and pretend transportation costs are zero, this form of assistance renders a person homeless.

Resellers like Wade were, in theory, eligible to participate in the program, but their business model, which involved flipping often-misrepresented trailers for tidy profits, left little incentive for them to do so. The refund program did not cover any resident that was not the primary purchaser, be they renters or secondary purchasers, even if they experienced sickness. Of the 10,839 units sold by January 2008 (Associated Press 2008), federal records I acquired indicate that fewer than one thousand were refunded.

In January 2010, a court order requiring FEMA to retain the trailers as evidence for a civil lawsuit expired. Large-scale markets lurched into motion as FEMA sought to liquidate its vast supply of toxic housing units. Instead of small-scale buyers like Wade nabbing fifty or a hundred trailers at a time, the new buyers snatched them up in tens of thousands, paying tens of millions of dollars. Auction houses like the Livingston, Louisiana–based Henderson Auctions invested over $18 million in FEMA trailers. Others, like Greenlawn Homes of Columbus, Ohio, paid $27.5 million for just over fifteen thousand units. These mass buyers deployed a battery of techniques for garnering profit.[4] Some of Greenlawn Homes' FEMA trailers were placed in mobile-home parks they owned, some were wholesaled, and some were dismantled for parts, their components used for individual trailers that could be sold anew. Greenlawn Homes sold units to manufactured home dealers, mobile-home park owners, and even auto dealers trying to use up idle space on their lot (Odendahl 2010).

Most trailers would be resold three to five times before reaching a residential destination (unlike the traditional mobile home that rarely moves after its initial placement). A few hundred or thousand dollars would be added to the sticker price with each change of hands. As the trailers were hauled to such far-flung locations as California, New York, and New Mexico, transportation costs ticked prices even higher. In this fashion, the multimillion-dollar arterial avenues of redistribution subdivided into more modest capillary circuits. Ultimately, approximately 150,000 FEMA trailers made their way from spectacular and worrying accumulations on staging areas to isolated and humdrum existences, perched on a mountainside in West Virginia, hauled onto a reservation in North Dakota, or planted in a highway-side RV park in Oklahoma.

Frenetic Space

I met Kathy, a white woman in late middle age, and Isabella, her white granddaughter, in a Walmart parking lot, a site for many of my rendezvous. It was July 2011, and we were in rural Indiana. Isabella was curled up asleep in the back of their gold minivan. Kathy rifled through documents. They wore matching pink outfits. We had first spoken several months earlier, when Kathy left a desperate comment on an online news story featuring research I had undertaken in Nebraska. On August 11, 2008, she and her husband, Cody, signed a deed to buy a FEMA mobile home for $27,000. The fifteen-year contract held a 12 percent interest rate, more than double the average mortgage rate at the time. The seller told them that one of the documents they were signing was a notice that the unit had tested clear of formaldehyde. This document turned out to be a notice of possible formaldehyde that, upon signing, indemnified the resellers from any future health consequences.

Kathy and Cody placed the home on a small parcel of rural land that they were also renting to own near the Ohio River. At the time that they purchased the trailer, Kathy was fifty-one; Cody, forty-seven; and Isabella was just born. Over the next four years, their health declined. Cody's colon cancer advanced, as did Kathy's kidney disease. Kathy began contracting bronchitis with regularity. Cody routinely came down with pneumonia. Cody's doctors put him on a continuous oxygen supply via nasal cannula after his blood-oxygen saturation decreased from 98 to 82 percent. Isabella developed intermittent nosebleeds, recurrent congestion, and gastrointestinal pain. They all suffered from headaches. Isabella's room tested at 71 ppb for formaldehyde. In June 2011, her physician signed a document I reviewed, stating that she "should not be housed in a formaldehyde-laden residence."

After twenty minutes of tailing Kathy's gold minivan, I pulled up to their home, tucked into a slight recess by the side of a sleepy road. Isabella was now awake and excited to entertain. She spun around gleefully—with outstretched arms—in the back of the minivan, announcing, "This could be my new FEMA home." Upon her doctor's recommendation, Isabella was sleeping at a neighbor's house. Since leaving the FEMA unit she was feeling better and only developed "ghost pains" when she was in trouble. Knowing the routine, she waited patiently out front while I went inside to meet Cody (fig. 1.6).

A simmering, frenetic energy engulfed Kathy almost immediately after she crossed the threshold. She delivered an urgent, almost feverish, accounting of home-wrought injury. She dug deep into the bottom of her bedroom closet, tossing lone shoes, old blankets, and children's toys out of the way. She beckoned me to join her in her burrow and pointed out several patches of persistent black mold along the baseboard. "It wasn't here when we moved in. It's probably from the river," she speculated. Cody watched from his recliner—he spent both his waking and sleeping hours there—oxygen tank at his side. I noted a photograph of teenage Cody in uniform and crew cut, standing by a US flag. Raising my voice to be heard over the TV, I asked if he was a career military man. He wasn't. After his time in the armed services, he had joined the Indiana State Police as a canine officer and trainer. "He trained Belgian Malinois. Trained 'em for finding drugs and search and rescue," Kathy hollered from the floor as she packed odds and ends back into the closet. "He could make them go to the bathroom on command," she recalled with a touch of nostalgic sadness. Cody affirmed with a small stoic nod. She marched across the house to get some bills ready for the post office as I sat down with Cody.

At several points in the conversation, Cody became exasperated when discussing Isabella's health and appeared to momentarily lose consciousness. His head would droop and his speech would pause, only to click back into gear, seemingly a sentence or two downstream of where he left off. I had trouble reconciling Cody's loss of somatic control with his former complete control not only of his own body but also of a canine accomplice. I was also developing a headache. When the air-conditioning engaged, the lights in the front room came on. "Oh, it does that, we're going to try to jerry-rig it," Cody remarked as an aside. Kathy pulled me into Isabella's room to witness the site of exposure with my own lungs. A neighbor appeared in the doorway. Kathy shifted to the right, inserting herself in the space between him and me. She spoke quietly through her teeth to me, "Don't say anything." After a few long moments of his presence being ignored, the man disappeared from the doorway, and Kathy's body language radiated relief. "He bought three of these FEMA homes and is trying to resell 'em," she explained. "We don't want to have anything to do with him. He's just trying to use our information."

FIGURE 1.6. Isabella playfully passes time outside her FEMA mobile home near Corydon, Indiana, July 2011. Photo by Nicholas Shapiro.

The atmosphere was charged with presumptive duplicity, excited by the false assurances of the "crooked" resellers and walls that expose. Deceit became a background noise of the domestic; the familiar was suffused with suspicion (fig. 1.7). Everything seemed to harbor the potential for, as Kathy put it, "trickery." Many FEMA home inhabitants, including Kathy, developed insomnia and got as little as two hours of sleep a night. This weariness only added to the second-guessing. Misgivings of everyday materials and people accrued on top of the American cultural practice of what Kathleen Stewart refers to as "scanning," a form of hypervigilance in which people "give shape to their everyday by mining it for something different or special" (2007: 39, see also Stewart 2005; Pine 2019). The lens through which Kathy and many other FEMA trailer residents read the world is defensive. It is not hopeful that this disposition will discover "something different" to inflect their lives with meaning; rather, this scanning hopes to apprehend and deflect the interruption of otherness into their lives. It is a precautionary measure to minimize inchoate domestic wounding.

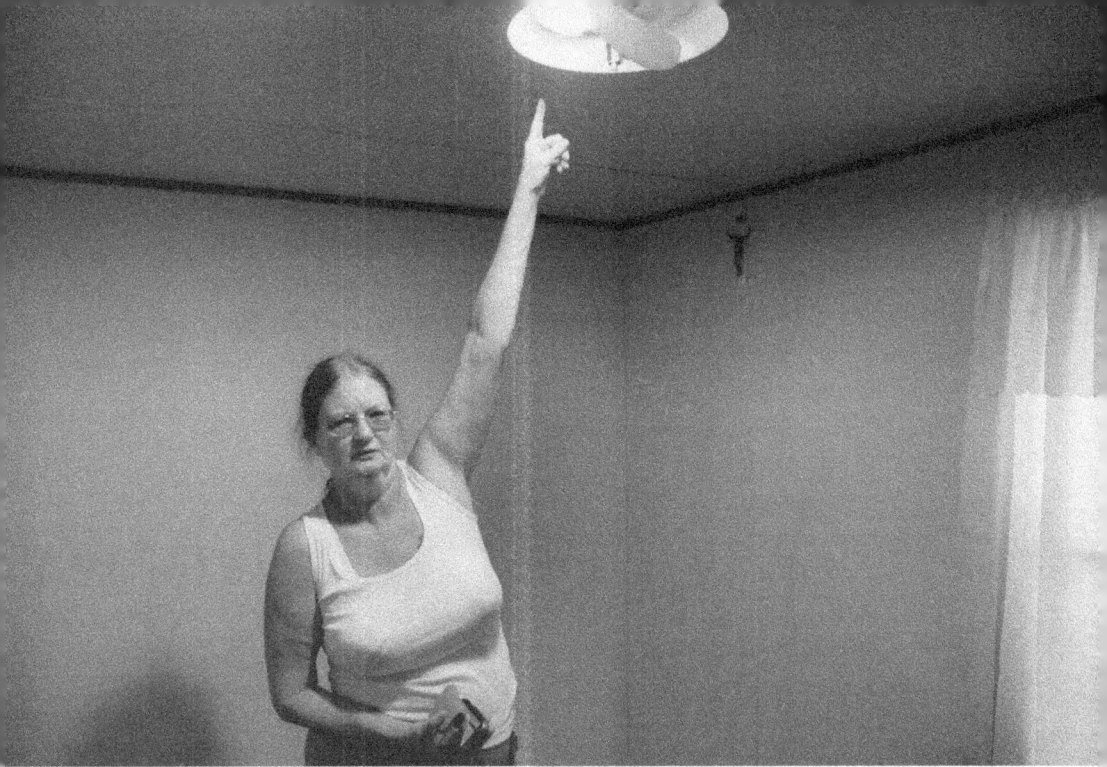

FIGURE 1.7. Kathy pointing to where she hung her formaldehyde self-test near Corydon, Indiana, July 2011. Photo by Nicholas Shapiro.

In the fall of 2012, Kathy and Cody secured a Veterans Affairs grant and bought a conventional home on a different five-acre lot, nestled in the woods. Kathy refers to it as her "dream home." While still wracked by chronic pulmonary issues, Kathy remained pleased with their move. She described the home to me in an email: "It is in a clearing in the woods, [. . .] a herd of turkeys were just in the back yard. It is a house not a mobile home. Cody now is only on oxygen at night instead of 24–7. That proves it was the mobile home."

Latent Suspicion

As trailer inhabitants' attempts to treat their emergent illnesses failed, they increasingly came to view their physical constitution in relation to the form and composition of their trailers. Having made this connection, many took to altering their homes in the hope of chemical respite and wellness. The most common measure was simply to keep windows and doors open as often as possible to dilute the indoor atmosphere of the trailer with, presumably cleaner, outdoor air.

Others, moved to exasperation by chronic illness, took more dramatic actions. An Ohio man cut out his floor and replaced it with hardwood flooring. Others replaced or removed cabinetry and interior doors made of formaldehyde-rich fiberboard. A divorced forty-nine-year-old Oklahoma woman and Cherokee citizen, Christa Perez, removed the trailer's built-in furniture, composed of engineered wood, and replaced it with solid wood furniture. She coated the remaining exposed wood with a "formaldehyde-free" polyurethane sealant to inhibit off-gassing. These home alterations never fully mitigated trailer residents' illnesses. Residents perceived little or, more often, no beneficial effect on their physical well-being.

Tests conducted by Lawrence Berkeley National Laboratory on the four most common FEMA trailer models revealed the range of FEMA trailer components that emit formaldehyde. Formaldehyde seeps from bed decks, bench seats, cabinets, carpets, the ceiling, the subfloor, seat cushions, walls, curtains, and interior doors. While there was great variability across models, walls and cabinets were the top contributors of formaldehyde to indoor air (Maddalena et al. 2008: 27–38). The omnipresence of components that leached formaldehyde into the indoor atmosphere made renovations both necessary and somewhat futile—especially for the chemically sensitized who can have severe reactions to even lower-level exposures.

Christa's symptoms, which included eye irritation, a burning respiratory tract, a new asthma diagnosis, fatigue, insomnia, general weakness, memory loss, and nightmares, did not go away after she altered her trailer. After she moved in, one of her cats began coughing and her Chihuahua developed skin allergies. "I worry about my pets," she told me during my three-day visit to Oklahoma City. "I leave, and they don't. I leave the door open when I'm here, and the air gets better, but when I'm at work or class I have to seal it up." Christa was concerned about the inexpressible ills of her animal companions and the potential future effects of the prolonged exposures they endured while she was at work or college.

Christa had moved into her FEMA home after quitting truck driving, taking up a part-time job at a charity, and enrolling in community college. She needed to reduce her overhead to get by. So, instead of moving, she flags ads selling FEMA trailers on Craigslist during breaks from long homework sessions. "It clears my head," she told me with a smirk. She was also one of the only resold FEMA trailer residents who expressed apprehension about the effects of the trailer on her *future* health. "I am worried about the long-term damage that it could have on me as a human being. It's pretty unknown," she told me over the phone, several months before I was able to meet her in person.

FIGURE 1.8. Christa's home and (mostly repaired) vehicle, Oklahoma City, Oklahoma, March 2012. Photo by Akasha Rabut.

Between the time of our phone conversations and when I visited Oklahoma, a semitruck ran a stop sign and struck her car (figs. 1.8 and 1.9). Both Christa and her daughter, who is grown and lives on her own, sustained substantial back injuries. Christa's concern with speculative chemical injuries was cut short, severed by the sharp pain of her spinal fractures and further diluted by the effects of prescribed pharmaceuticals. As she joked to me over email, "Time doesn't matter much when you're on pain pills and muscle relaxers. LOL."

Long-term risk is not a fully rendered figure for those who came to inhabit resold FEMA trailers. Only four other interlocutors worried out loud about the long-term effects of domestic exposure on human health, and they directed such concern exclusively toward their children. Indicative of the potentially classed nature of having the time to worry about future unknown harms, three

FIGURE 1.9. Christa's home and (mostly repaired) vehicle, Oklahoma City, Oklahoma, March 2012. Photo by Akasha Rabut.

of the four used their FEMA trailers primarily as recreational vehicles, very uncommon use among those that contacted me. The other was a white, exurban FEMA mobile-home resident in Alabama. The suffering of durative chemical exposures and defective domestic infrastructure conjures not a forward-looking weariness to coming interruptions but suspicion of an ongoing present, studded with both expected and unexpected dysfunction and nearly habitual enervation.

CORROSIVE HAPPENINGS PATTERN the toxic home's domesticity in ways that appear meaningless, puzzling, or enfeebling. They conjure a double-dealing way of knowing and theorizing how to change one's circumstances, knowledge that can easily drive someone mad. Vast differences in how bodies react to chemical-laden atmospheres can drive a wedge between the acute surreality encountered by one person and the conventional reality experienced by the person lying in the bed next to them.

The surreal state of things produced by the FEMA trailer, or the toxic home in general, is not unique. Rather, the wounds of late industrial life tend toward surreal bruising and warping rather than direct lethality. Think of the altered trill of songbirds exposed to polychlorinated biphenyls (PCBs) (DeLeon et al. 2013); dunes engulfing Chinese towns in large-scale, slow-motion desertification (Zee 2017); tangles of bright plastics found within the carcasses of albatrosses in remote Pacific atolls that, surprisingly, are not their cause of death (Liboiron 2021); or the way rising CO_2 levels can sap varieties of cereals or legumes of vital nutrients like zinc, iron, and protein, making the same meals gradually more anemic (Myers et al. 2014). Think of the small ways that the massive geophysical destabilization of climate change inflects ordinary scenes with difference—as innocuous and yet unnerving as a fly buzzing in the US Midwest at Christmas (L. Moore 2010).

Gleaning meaning from illogic and dysfunction is fundamentally difficult. Yet, all across the archipelago of toxic homes, the chemical reactions that are wilting bodies also catalyze investigations and agitation. Building up understanding by being worn away. Irruptions of illogic are aggregated in the minds and bodies of the exposed. They leverage that domain to which we all have unique and sometimes lonesome access—our own somatic experience—to anchor their sense that something is wrong in their environment and orient how to intervene (cf. Kenner 2018).

TWO

From Chemical Fetishes to the
Late Industrial Sublime

Relief swelled up from somewhere deep in Lizzie's chest when Federal Emergency Management Agency (FEMA) contractors arrived to install a trailer next to the concrete slab where her home had once stood. With little bursts of excitement, Lizzie celebrated her small emergency housing unit. It had arrived some six months after Hurricane Katrina had splintered her house in Pass Christian, Mississippi. Of note was its "bump-out" panel that considerably expanded the floorspace, its clean-looking interior, and, most tellingly, its fresh smell. "I love the new-car smell—why, everybody does that has air fresheners, they smell like that, ya know. And that's what it smelled like to me. And I was like, 'YAY!! They brought us a brand-spanking-new one!!' I thought they would bring us an old ratty one, ya know?" To Lizzie, and to many others, the trailer's chemical bouquet appeared to reveal the essential and auspicious quality of the structure.

"I was really happy with it," she recounted to me five years later, sitting on the living room floor of her new home in a suburb of Memphis, Tennessee, near where she had grown up. It's also near where she, her two children, and her boyfriend, all white, had stayed with family in the months after the storm—living in an extended holding pattern while cooped up in borrowed space. Letting her fingers sink deep into the carpet, Lizzie recounted her short-lived fondness for life in the trailer, "to have to stay somewhere for free, it was great!" before continuing in a hushed tone, "...at first." Lizzie, delighted with her family's newfound privacy and sense of normalcy, quickly became pregnant with her third child. The "new-car smell" that so excited her and that seemed to indicate the trailer's quality when she and her family first moved in quickly became overwhelming with her heightened scent perception during pregnancy. She vomited every time she entered the trailer, becoming dehydrated to the point of hospitalization. Her son simultaneously developed asthma and severe eczema. "His skin just started peeling off in there," she stoically remembers. Like his mother, he also started spontaneously throwing up. Unlike her, he didn't stop vomiting when they left the trailer. For years he averaged three to four upchucks a week, often resulting in school sending him home early. From time to time, Lizzie would hear a small guttural noise coming from the backseat of her minivan and catch a glimpse of him in the rearview mirror, stealthily puking into a large cup. These little eruptions, once urgent signs of some latent disturbance—of a surreality that was both disorienting and demanded urgent remedy—slowly became ingrained in the rhythms of everyday life. Six months after moving into the first trailer, they relocated to another, less pungent, FEMA trailer. Her symptoms abated within two weeks, while her son's illnesses persisted in a diminished form.

Lizzie is not alone in having initially associated the bevy of volatile organic chemicals (VOCs) that have come to be known as the "new-car smell" or simply the "new smell" with value or freshness. "It looked clean and had that new smell. I was pretty happily surprised," an elderly woman in suburban Tennessee noted with satisfaction after purchasing a secondhand trailer, before she developed simmering health problems.[1] The molecular makeup of that smell had pulled her into purchasing the home, and it later drove her out of it. Again and again, in Michigan, Illinois, Florida, and West Virginia, residents of homes that would eventually be suspected of harboring elevated formaldehyde levels told me that the "new smell" imbued chronic toxic exposures with pleasure, fashioning the aroma as a selling point.

This chapter begins with a forensic genealogy of the "new smell" as a fetish of the late industrial sensorial palette. This exposure experience, wherein somatic perception has been commodified to the point that it is desired, represents an ultimate form of chemical-capital surrealism. The olfactory pleasure of chemical harm has become so enmeshed in the late industrial sensorium that it has simply become reality. The chapter then seeks to understand how exposed people can cut through a surreal sensibility crafted by corporate advertisers. How can hazardous exposures, so normalized that they are desired, be reckoned with to open up the pursuit of change? One answer emerges from the less nameable aspects of the body's encounter with the environment. Chemical encounters subtly and corrosively alter the bodies of those living in homes with compromised indoor air quality in sometimes subtle and striking ways.

Researchers have primarily discussed the "discovery" of invisible chemicals via scents and science. Here, I draw our attention to another way of knowing, the attuned body, as the primary substrate of domestic formaldehyde exposure discovery.[2] Bodies are sensors that indicate the presence of toxicants and, in some cases, specify their atmospheric concentration with uncanny precision. The empirical matter that fills this chapter is intended to first challenge and then expand the confidence that we often place in our own ability to know when we have sensed something and when we have not.

Human bodies reveal imperceptible chemical exposures with their own, often subclinical, wounding. In these protracted encounters, the somatic precedes and then is entangled with the rational, a mingling of mind and body that bucks the all-too-common dismissal of domestic chemical complaints as psychosomatic. I'll argue that these affective processes of attending to the minute aberrations of the body and atmosphere are our primary means for discerning protracted and low-level encounters with domestic chemicals. This way of knowing by tracking small changes to body and atmosphere across time and space can accumulate into a process I call the "late industrial sublime," which elevates minor enfeebling encounters into events that stir ethical consideration and potential intervention.

An Intoxicating Aroma

Many of the people who mentioned their fondness for the new smell conceded that the chemical aroma itself was not fully enjoyable, as it "burned a little." Nevertheless, for Lizzie, and others, the social currency of the smell overrode the raw nature of its sensory experience. Similar to "the tingle means it's working" ad campaign of the dandruff shampoo Denorex in the 1980s, the slight singeing sensation conjured by the chemical brew known as the new-car smell signals a modern genre of functionality. It, also, is the primary idiom through which people living in toxic homes expressed an awareness of indoor air.[3] The siren song of "new" chemical brews wafts into the lives of consumers only for those alluring encounters to melt into the surreality of everyday, ongoing intoxication. To understand how markets normalize chemical exposure and how now-familiar toxic encounters might be called into question, I ask: How did this scent concept, which commandeers bodily knowledge of exposure and repackages it as the sensation of freshness, become almost universally recognized in the United States as an index of newness for virtually all commodities with synthetic components?

The new-car smell entered the American sensorial lexicon in the late 1940s. Born out of the mid-twentieth-century US auto industry, the cultural saturation of the new-car smell is tied to the twin rise of automobility and the proliferation of petrochemical products. As World War II ended, America's appetite for automobiles soared. Car enthusiasts were near feverish by the time surrender documents were signed in Europe and the Pacific, as the war effort had halted the production of civilian passenger cars in February 1942. During this period, the olfactory pleasures of a new car began to percolate through the written record before eventually coalescing into the scent concept of the "new-car smell."

A 1945 issue of *Popular Mechanics* predicted that the release of 1946 edition cars would likely be "as exciting to most Americans as the first transatlantic flight."[4] The magazine located the sensual pleasure of automobile ownership not only in its distinctive smell near the time of purchase but also in the mechanical neighing of its engine: "Americans are extremely eager to hear that musical squeak of a new automobile and to smell the fresh enamel and hot metal."[5] The giddy prose of the moment immediately after the war, which interpreted the din of midcentury engines as enjoyable, quickly sobered up. A year later, a 1946 motor-oil ad in *LIFE* magazine depicts two men in suits admiring a pristine car. The copy reads, "There's something about the very smell of a new car that gives you a big thrill." The ad emphasizes "the clean odor of new upholstery" and teeters on the idiomatic precipice of the new-car smell without naming it as such.[6] By 1948 the scent concept had solidified, as evinced in an advertisement in the June 11 issue of the *US News and World Report*: "You open a door and sniff that 'new-car smell' of fresh lacquer and newly-loomed upholstery. From beneath the gleaming hood a faint whispering speaks eloquently of the torrents of power eagerly waiting to take you anywhere on the map."

The "musical squeaks" of three years before had been subdued to "faint whispers," while the smells of "fresh enamel and hot metal" that ambiguously spanned the cab, the exterior, and under the hood crystallized into a nameable sensorial experience specifically within the passenger compartment. The new-car smell was groomed into the dominant aesthetic experience of life behind the wheel (fig. 2.1).

The plastics, adhesives, sealants, synthetic upholstery, and lacquered wood panels that conditioned the air inside post–World War II cars constituted both a new domain of heightened chemical exposure and a new market for the pet-

Great Day

FIGURE 2.1. *Left:* Open-hood image from an advertisement in *LIFE*, 1946. *Right:* Open-door image from an advertisement in *US News and World Report*, 1948.

rochemical industry. At midcentury, the average auto consumed approximately ten pounds of antiknock compounds and a nearly equal amount of lubricants and other petrochemical fluids per year. Automatic transmissions, introduced in the late 1940s and early 1950s, also increased chemical profits, as they used approximately ten times the volume of lubricants and chemical additives as standard transmissions. These inglorious applications under the hood quietly ensured that the power and convenience of automobilization did not grind to a metal-on-metal halt.

Petrochemical products were key to provisioning the futuristic and privileged experience in the cab. This much was made clear in a 1960s issue of Dow Chemical's trade journal, the *Dow Diamond*. The article, which describes the bonanza of applications for new synthetic fibers in automobile interiors, almost exactly parallels the attentional focus and gender dynamics shown in figure 2.1, "Back in the forties a man would saunter into his favorite showroom, and the very first thing he'd ask the salesman to do would be to unlatch the hood and expose the motor." The article continued in typical midcentury the-future-is-now style: "Today, the cast has an additional character, the emphasis has changed. Now the American wife joins her husband in the local dealer's showroom. . . . In an instant her eye critically appraises the fabric on the seats, the door panels, the headliner" (fig. 2.2). In Dow's narrative, the essentialized eye of a woman and the figure of the acquiescent husband that "agrees with her, nine times out of ten," on issues of aesthetics provides

FIGURE 2.2. A photo from the *Dow Diamond* features fabric using Lurex, a fiber with a metallic appearance made by sandwiching a thin filament of aluminum between plastic films.

justification, without troubling normative gender roles, for shifting focus from the torque and utility of the engine to the status and beauty of the interior.[7]

After the war, designers bested engineers in long-standing struggles for influence within the US auto industry. Competing through style proved to be less expensive for GM, Ford, and Chrysler than price jockeying or engineering innovations. As a result, stylists "unilaterally dictated auto designs, with engineers reduced to trying to make their increasingly outrageous dream machines run" (Gartman 2004). While the silhouettes of sleek auto bodies rightly garner prominence in the historical memory of this fantastical moment of car design, the large-scale overhaul of the interior—made possible by combining newly available petrochemical products—proved to be much cheaper than modifying the exterior. This confluence of forces—massive economic and consumer expansion, a proliferation of alluring and off-gassing petrochemical materials, and a spike in marketing the sensuous pleasures of a car's interior—generated the sense that the new-car smell was a unique sensation reserved for the privilege of new car owners. These heightened spaces of chemical exposure quickly pervaded the United States during the "frenzied" car sales of the 1950s (Offer 1996). The written record indicates that auto industry advertisements were largely the first to mention the new-car smell, associating the miasma with status, style, cleanliness, and freshness.

By the mid-1950s the new-car smell had become a readily known sensorial reality that sprang from advertisements into the pages of fiction. The main character of a mid-decade suspense novel mentions not just the smell itself, but its manipulative use as a means of synthesizing newness: "You know, Gina, in the used car lots when they get a good clean car in, a recent model, they sometimes brush the upholstery with embalming fluid. That gives it a new car smell. Psychological salesmanship" (J. MacDonald 1955: 150). The embalming fluid, presumably formaldehyde, was used to preserve the most fleeting aspects of the car's aesthetic newness: its aroma. Smell is the ultimate test of time. Even if an automobile is not driven and is kept in pristine condition, the scent of its chemical nascence will slowly fade when it leaves the factory floor. Like a populist radiocarbon dating, consumers essentially tested chemical decay rates with their noses as a proxy for age.

Over the intervening half-century, the new-car smell gradually moved from being referenced as simply a joyful "free" add-on of a newly purchased automobile to being a stand-in for the act of buying a new car. From headlines of the *New York Times* (Maynard 2009) to YouTube video accounts of driving the latest model off the lot, the new-car smell is synecdochic for the new car in its entirety.[8] Some contemporary advertisements label the aroma "the smell of success," an automotive analog to the "smell of money." The smell is so canonical that in 1995 Crayola added a "new car" crayon to its Magic Scents collection, alongside daffodil, cedar chest, and eucalyptus. Synesthetically advancing the pristine associations of the smell, the crayon was blue. The two other blue crayons in the set bore the healthy names of soap and fresh air.

The fragrance is steeped with so much cultural capital that it became just that: a perfume. Responding to research that demonstrated a consumer preference for the hyperreal (Baudrillard 1995) smell of artificial leather to that of real leather, Ford began infusing its cars with a chemically concocted scent at the turn of the twenty-first century. Cadillac followed suit with its own scent, named Nuance, three years later (Lindström 2005: 94). Perfumists closed the associational loop between male bodies and cars. Mustang offers a brand of cologne, available at Walmart. Ford Spain released its own bottled fragrance, dubbed Olor a Nuevo ("smells new") in the spring of 2012, which was sold individually as well as infused into Ford's certified pre-owned vehicles. A contract guaranteeing the smell was added to the sales documents for these used cars. Newspaper classified sections recognized the distinction of these used vehicles by recategorizing them in a novel grouping between pre-owned and brand-new: "Usados con Olor a Nuevo."[9]

While the chemical scent was infused with excitement in North America and some parts of Europe, the smell is by no means universally appealing. In China, the new-car smell is more odor than fragrance. In China, unpleasant odors were routinely the most frequent complaint of new car owners and a top reason for not buying a car—surpassing fuel economy, engine performance, and safety. Some consumers noted the potential health risks associated with these smells (J.D. Power 2017). As the material sources of these VOC emissions are too ingrained for many manufacturers to extinguish them at the source, Ford applied for a patent in 2017 for an autonomous vehicle odor-remediation protocol, in which cars drive themselves, unoccupied, to a sunny spot and rely on solar heat to bake the chemicals out of the car.

Fresh to Death

The new-car smell became a sensation amid post–World War II Fordist regimes of mass production and mass consumption. At midcentury, this automotive techno-optimism ushered in "a specific mode of living and of thinking and feeling life" (Gramsci [1947] 1991: 597). The chemical bouquet of the new-car smell became a highly pervasive, yet still ambiguous, feature of this emergent aesthetic and sensorial palette. This olfactory side effect of mass production became a product in its own right. Spilling into international markets, and branching out past the auto industry, the hodgepodge of VOCs that came to be known as the new-car smell has become the scent signifier of newness for a broad range of products: from new planes and houses to new electronics, carpets, athletic wear, and this very book (in its hardcover or paperback form, at least) (fig. 2.3). What gets dubbed as a "new" smell likely has slightly different chemical and perceptual qualities across these varied contexts, but it shares a common meaning.[10]

Repeatedly, manufactured-housing inhabitants would respond to my asking if their home had any particular smells with a thoughtful pause followed by a resolute "No." Yet, when I followed up by asking if their used trailer had a new-car smell, they would quickly respond in the affirmative. Looking back, it appears that they assumed that by "smell" I meant "smelly," an out-of-place or unpleasant scent—a category antithetical to the desirable new smell. Time and

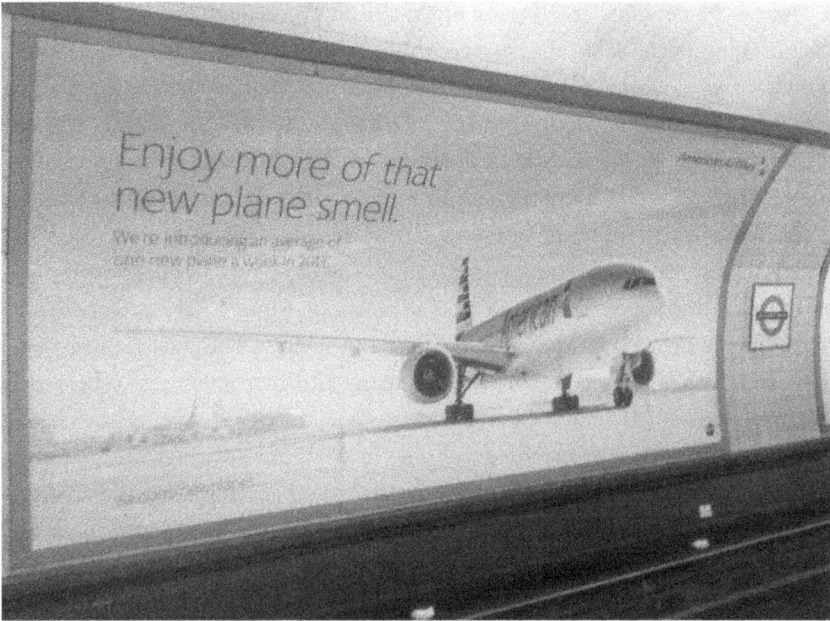

FIGURE 2.3. American Airlines advertisement in the London Tube, 2013. Image by Nicholas Shapiro.

time again, they invoked the "new smell" as their first observation and primary indicator of the FEMA trailer's quality. The scent suggests a newness of essence that transcends appearance.

Aided by marketing that began by protecting one product and then proliferated a whole range of spin-off products, the new smell has become a popular practice in North America and Europe for embracing chemical exposure without naming it as such, the perceived value of the smell obscuring its molecular hazard. In line with Karl Marx's notion of commodity fetishism, fixation on the value of a new smell masks the chemical exposure that it also indicates. It "transcends sensuousness" (Marx [1867] 1990: 163), belying its chemical nature and its potentially toxic effects on the body.

In its more pervasive form—as an accidental olfactory addition to synthetic products, rather than as a product of its own—the new smell is not a commodity. In the form that is inhaled with a sense of contentment in new and refurbished homes across the country, the smell is an ethereal reassurance that the value of the home has not diminished over time. While not a commodity in the strict sense, the new smell remains a fetish that conceals its origin and substance.

Access to the new smell does not uniformly percolate through America. Manufacturers and advertisers target those excluded by whiteness and capital, who have never come close to owning a new car, much less a home, for desiring exposure under racial capitalism (cf. J. Jackson 1994: 61). Sprays that provide a new-car scent synthesize the rarified, yet unattainable, sensations of driving a new car off the lot. When purchasers of used trailers took pride in the smell, it was often in the context of homeownership having been financially unthinkable before stumbling on their cut-rate unit. The smell was part of the trappings of their triumph and newfound status. Inversely, when I have presented versions of this chapter in academic settings and handed out (low- to nontoxic) samplers of synthesized new-car scent, many audience members vehemently announced disgust elicited by the aroma. Their reactions were indicative of a counter-reading of similar molecules, one that sought to distinguish their taste, their ability to discern freshness from something that could be purchased in a plastic wrapper at the dollar store. The reactions elicited by the new smell perform a classed and ambiguously raced sensibility.[11]

Not long after John Updike described the "sweet tangy plastic new-car smell" (1960: 107) as a reassuring but fleeting moment of transcendence, the petrochemical components that give rise to the new car became key figures of countercultural critique. For the hippies on the West Coast and the New Leftists on the East Coast in the 1960s, plastic became a metaphor for society's ills. As Jeffrey Meikle has noted, in these circles plastic was commonly used as a term of disdain for the "artificial," "meaningless," "rigid," and "materialistic" "white honky culture" (1995: 260) that emerged after World War II and at the dawn of the Cold War. Tracking exactly how dissident petrochemical aesthetics from the 1960s braided together with the turn-of-the-millennium mainstreaming of organics, wellness marketing, subsumption of grunge fashion into prestressed clothing, and the commodification of authenticity to form countervailing perceptions of the new smell would be a project of its own.

The specter of the automobile and the vast petrochemical infrastructures that make it possible haunt the American sensorium—the very foundation of perception. Whether adored or abhorred, the history of the new smell complicates assumptions of smell's capacity to apprehend invisible chemical hazards. Alongside this fetishized scent of exposure, much more diffuse sensory practices register their chemical encounters in other subtle and yet exacting and powerful ways.

This has led me to wonder: How might expanding the avenues and temporality of sensing yield an appreciation of what many of us are abbreviating from our own sense of the world?

Attuning to Distributed Harms

A three-year-old staying in a travel trailer in rural Oklahoma developed small red dots on the backs of her ears, began bruising more easily, and walked through the world more clumsily, constantly toppling over. An older Indiana woman descended into a fog of "fuzzy thinking" and felt that her body was deteriorating at a slightly accelerated rate. A middle-aged man in Ohio sustained a "sick stuffy nose" and "throat problems" for a year and a half. Eye and respiratory-tract irritation, headaches, insomnia, and fatigue slowly crept into the body of a single father in rural Florida. His dreams, which became increasingly menacing over a matter of months, abated in intensity only when he slept next door at his grandparents' house. The stool of a nurse in Texas gradually loosened in consistency. A Native police officer on a reservation in the Pacific Northwest almost entirely ceased eating as his sense of taste began to dull. His wife, experiencing the same sensorial skewing, doused her food with large quantities of salt and noted the oxygenic surreality of the "weird air" in their home. The people above, and those who fill the rest of this chapter, began experiencing subtle and ongoing alterations to their physical constitution after spending time in a home that would eventually be suspected of sustaining elevated levels of airborne formaldehyde. This is not just a FEMA trailer phenomenon but one that I observed across a range of toxic homes.

These slight biochemical impressions, which at first appear simply meaningless, puzzling, or surreal, accumulate in the bodies of the exposed and reorient them to the molecular constituents of the air and the domestic infrastructure from which such chemicals emanate. Through the articulation of these small corrosive happenings, residents overcome the pleasures of the new-car smell and reckon with how their homes are decomposing into them as they decompose in their homes.

The somatic work of the chemically concerned is enmeshed with an apprehension of their own bodies that is simultaneously sensuous and epistemological. The instrument of these lived experiments is the body. They occur within processes of bodily reasoning and yield bodily knowledge beyond the commodified allure of "new" smells. Over time, the information accrued by way of bodily reasoning accumulates to the point that the injured knower is able to articulate how tiny molecules deep within their or their loved one's body are connected to much larger problems. I think of this apprehension of indistinct harms that exceed conventional observation, atomized and distributed across countless homes, as a "late industrial sublime." Exposed people wield knowledge from their wounding to provoke reflection, disquiet, and contestation, despite the conventionally insensible nature of the molecule-by-molecule violence. I align this observed practice with the chronotope of the late industrial (Fortun 2012) because of the insidious distributed harms that mark this ongoing moment of deteriorating sociotechnical systems and economic, climatic, and infrastructural instability that is the inverse of Enlightenment-era conceptions of the sublime that serve the status quo.[12] The dissident potential of the late industrial sublime, with its ability to make meaning from inchoate and surreal happenings, is particularly well suited to reckoning with the large-scale hazards distributed by contemporary infrastructures of living under racial capitalism.

This practice of bodily attunement is not confined to those afflicted by multiple chemical sensitivity or those with diagnosed pathophysiologies, like asthma. Rather, these molecular and relational appreciations arise from a somatic susceptibility and epistemic capacity common to human life—and often informed by nonhuman life.[13] By definition, toxics bear "a potency that can directly implicate the vulnerability of a living body" (Chen 2012: 203). It is by virtue of this very capacity to be chemically wounded, even minutely so, that bodies bear their revelatory power.

Body Meter

In February 2011, Linda Kincaid emailed me in response to a call for participants in my study of the experiences of domestic chemical exposure. An environmental activist had forwarded the call to what she refers to as her "formaldehyde list." The list comprises a broad array of individuals interested in formaldehyde, many of whom have personally felt its effects—from former FEMA trailer residents, to consumers concerned about the broad range of products made with formaldehyde, to, evidently, industrial hygienists. Linda has worked as an industrial hygienist—a scientific profession charged with the responsibility of assessing, controlling, and communicating environmental hazards—since 1991. She holds a master's degree in public health from the University of California, Berkeley. The immediacy of her interest in domestic formaldehyde was derived not only from the elevated chemical levels registered by the monitoring equipment she had placed in the homes of her residential clients, but also from her own symptoms of exposure, which maintained a grip on her after returning from the field.

Before we met in person in suburban Los Angeles, where I attended one of her formaldehyde home inspections and learned to use a real-time formaldehyde meter, we spoke at length on the phone. Linda had only become interested in domestic formaldehyde exposure in the past few years. When she received her first phone call from a family that suspected their home was making them sick, she reacted with skepticism. "What are you talking about?" A quick literature review, however, soon revealed the possibility that common domestic formaldehyde levels could cause the symptoms this potential client reported. Linda's attention was piqued, and she began to amass a small arsenal of portable real-time formaldehyde meters. Yet the vast majority of her work continued to be for the semiconductor and solar industries. Even so, Linda's curiosity about the magnitude of domestic chemical contamination continued to bloom.

Developers swiftly rejected her offers to test new subdivisions at no cost to them. Not one to be so easily stymied, she saw clandestine testing of open houses as her best option for gauging the prevalence of elevated formaldehyde in residential spaces. On free weekends, she set out to tour and test new unoccupied homes by herself, with the intake hose of her Interscan 4160 formaldehyde meter timidly cresting the lip of her purse:[14]

> It was really kind of a lark. Can I find elevated formaldehyde in homes? Is it going to be one in ten? . . . Within a few weeks, I came to realize that there was a problem here. There is a *huge* problem here. I was getting the kinds of concentrations that they found in the FEMA trailers, and these are not trailers; these are high-end Silicon Valley homes.
>
> And I started noticing that homes in one city in particular had seriously raised formaldehyde as compared to others. . . . Every house I went into had really pretty high formaldehyde, and I would have a headache and have trouble sleeping that night and toss and turn all night long. I'd be exhausted the next day, and when I did other communities it seemed that the formaldehyde wasn't as high and I didn't have those responses to the same degree or maybe not at all.

As Linda began to log higher levels with her formaldehyde meter, she also began to log increased levels within her body. Her symptoms signaled elevated chemical levels as clearly as the LCD readouts of her assessment technologies. In embodying the invisible gas, she did not utilize the standard human sensory faculties but rather a calibrated, yet diffuse, awareness to aberration. She attuned to the irregular physical state of her neurochemistry.

Appraisals of her clients' homes would often turn back to her own body. When I asked about the curious symptom of intensified dreams that her clients reported, her first reaction was to describe her own corroborating experience: "And those were one of my symptoms too; it doesn't seem to happen to everybody. It absolutely is one of my symptoms. It is guaranteed. If I am in a house with 50–70 ppb [parts per billion] formaldehyde, I will have utterly weird, bizarre, freaky terrifying nightmares, and that is very consistent. It is not something that happens to me normally, so when it does happen it really stands out."[15] Linda's symptoms can emerge after merely an hour of exposure, bearing corporeal witness to long-term, low-level chemical exposure disorders that biomedicine has historically disqualified as (female) psychogenic illness (Murphy 2006). Her repeated experiences, in combination with her monitoring equipment, lend credence to individual and isolated complaints on the scale of reproducible and scientifically observed phenomena. *It is guaranteed.*

Despite the short duration of Linda's exposures, she can surmise formaldehyde levels with extreme precision. In the above quotation, she asserts that she can sequence the onset of exposure symptoms down to a margin of error of about 20 ppb. In liquid terms, that is roughly equivalent to determining the difference between fifty and seventy drops of formaldehyde diluted in a small railroad tanker or 250 chemical drums. In temporal terms, such accuracy is comparable to a margin of error of a minute when measuring durations over the course of a century.

At first blush, the exactitude of her body-meter-air attunement appears to border on the uncanny, if not the impossible. But the ability to discern such infinitesimally small differences in atmospheric concentration does not derive from a supernatural capacity on Linda's part. Rather, such perceptivity results from a mundane monitoring of both repeated bodily irregularities and the levels of formaldehyde found by her meter. These practices are born out of standard scientific method, professional curiosity, everyday corporeal awareness, and openness to being affected. Linda's embodied awareness of biochemical aberration does not lie beyond the realm of toxicological plausibility.[16] As "the exact mechanism of action of formaldehyde toxicity is not clear" (ATSDR 2014: 5), the aspect of this process that remains inexplicable relates to the limits of toxicological knowledge, not a mythic extrasensory perception.

The Late Industrial Sublime

Operating in tandem with her real-time formaldehyde meters, Linda's body viscerally logged the chemical exposures of the houses she visited. Over time, she calibrated an understanding of toxic effects to the outputs of her instrumentation, a process of indwelling both the indoor atmosphere and the meter. Scientific instruments and soma evaluated their immediate surroundings in accord. Linda's bodily attunement to the toxic effluvia of engineered wood architecture crept into her consciousness at a snail's pace and oriented her not only to individual emission sources but to the social and material patterning of exposure, a process of aggregating bodily knowledge that I refer to as the late industrial sublime.

Formaldehyde's presence in domestic space was not signaled by overwhelming sensory stimuli, as is the case in traditional invocations of the sublime. Rather, it is indicated by a thickening veil of indistinction as the observer's perceptual faculties become occluded. For the chronically exposed, the biological interference of air-quality-induced illness is received as a phenomenological transmission in its own right (Fortun 2003: 186). The sensorial noise of manifold irritations and illnesses is the signal of domestic chemical exposure and the bodywork employed to apprehend the qualities not only of indoor air, but also of other invisible distributed violences.

Since at least the dawn of the Enlightenment, the concept of the sublime has brokered philosophical relations between environmental exposure and the political status quo. For European Enlightenment thinkers, the concept of sublimation—the elevation of the grotesque or overwhelming grandeur into the control, delight, and superiority of reason—was part of a larger project of metaphysical optimism (Ray 2004: 10). For these thinkers, who united under the conservative credo of "whatever is, is right," the sublime could transform the terror of natural disasters or other horrors of the natural world into opportunities for reassuring oneself that the ultimate mastery of human reasoning was apart from and above the physical world.

For these men, the Great Lisbon Earthquake of 1755, for example, could be understood not as horror that killed upward of forty thousand people, but as an opportunity for learned men to reflect on their intellectual superiority to dominate the threats of the material world. But this isn't only an approach to thinking about exposure in the days of writing by candlelight. More recently, as what could be considered the peak of industrial modernity, anthropologist Joseph Masco (2004) has noted that some male nuclear weapons scientists who were knocked to the ground, flash-blinded, or otherwise pummeled by the world's first mushroom cloud felt a similar form of the sublime as they reveled in a feeling that approached divinity.

The late industrial sublime, which has been taught to me by dozens of chemically exposed people, is the opposite of the Enlightenment sublime and its atomic offspring in the form of the nuclear sublime. If the traditional sublime quells humanity's feelings of being dwarfed by the material world by providing reassurance of the power of the mind, the late industrial sublime constitutes a sensuous reasoning that indicates how open our bodies are and amplifies— rather than extinguishes—the tensions, agitations, and dissident potentiality of large-scale hazards. Absorbing minute quantities of toxicants over an extended period of time is not a situation that can be transcended by way of a feeling of rational control. It is the coalescing of underrecognized disturbances. It is the beginning of a confrontation, not its resolution.

As common industrial chemicals warp, distort, and decay human and non-human bodies alike, they corrode the optimism of the Enlightenment. Instead of "transforming the worst into the best" (Lyotard [1983] 1988: 41) as a foil of human triumph, the late industrial sublime condenses vaporous displeasures. It is a way of being deeply moved by the latent toxicity of industrial human progress.

For Linda, the prevalence of elevated formaldehyde gradually accumulated into a technical and embodied awareness of residential chemical exposure that dwarfed her by its scale. *Within a few weeks, I came to realize that there was a problem here. There is a* huge *problem here.* This profound apprehension stemmed from widespread private, indistinct, chronic, and fragmented phenomena. Again, this sharply contrasted with Enlightenment ideas of the sublime as public, spectacular, and overtly violent.

If bodily attunement is the dynamic process through which knowledge of individual spaces of chronic exposure is somatically attained, the late industrial sublime is the accrual of information by the attuned body to the point of articulating the patterned practices and infrastructures that distribute pockets of exposure across space. It is the traversing of a threshold of chemical awareness whereby the *irritations* of one's immediate environment become political *agitations* to apprehend and attenuate the effects of vast toxic infrastructures. This apprehension then often demands additional forms of labor.

Linda approached the city council of San Jose, California, in the summer of 2009 as its members were on the verge of passing a building ordinance that required new homes to be certified as "green." The bill's energy-efficiency requirements would likely mean that homes would be sealed more tightly, a measure that would, in turn, likely result in higher domestic formaldehyde levels.[17] Linda proposed an addendum requiring green homes to be tested for and meet indoor air-quality standards. The formaldehyde levels logged by Linda's instrumentation were already well in excess of government-recommended thresholds. She offered to render her services at no charge to the city or developers to demonstrate that she held no financial conflicts of interest. Her proposal was met by a smear campaign financed by the Formaldehyde Council, an industry-funded interest group, which commissioned scientific assaults on her findings. Linda was warned that her assertions about widespread domestic toxicity put her "at risk for future litigation," as systems of commercial asset protection transformed her effort to mitigate systemic exposure risks into legal, scientific, and financial risks on an individual level.[18] Her data were then ignored, and her motion scrapped.

Articulating constellations of manufactured hazard was threatening enough to the engineered woods markets that a multinational association was motivated to bring the power of its law, science, and capital to bear on the proceedings of one of the smallest scales of governance. Even though this one instance of political action was quickly bullied away, the potential for pursuing change as a result of bodily knowledge gained from conventionally invisible harms far exceeds the potential of industry lawyers to squash it.

Bodies of Evidence

Linda Kincaid's bodily attunement grew out of professional and personal curiosity. More often than not, however, bodily knowledge of chemical exposure derives from the necessity of cohabiting with toxins, as was the case for Harriett McFeely and her husband, Dick. The couple live in a modular home on the outskirts of a small town in Nebraska. In the spring of 2011, I traveled to stay and speak with the McFeelys, who claim to have endured more than two decades of domestic formaldehyde exposure.

Before Harriett got access to free formaldehyde tests from the Sierra Club and before formaldehyde had been introduced to her as a possible perpetrator, she was near the end of her rope. During her twenty years of inhabitation, she had slowly developed constant diarrhea, a runny nose, fatigue, severe eye irritation, double (and occasionally triple) vision, headaches, a sense of taste that skewed toward metallic or simply "strange," and numerous other symptoms.[19] With resurgent exasperation she recounted her dogs getting sick and dying (fig. 2.4), one after the other, while her and her husband's health steadily deteriorated. Her doctor received her complaints with skepticism and an implied diagnosis of hypochondria: "They couldn't find out what's wrong in my body, so they thought I was crazy. That's the only answer."

Harriett first began suspecting the house as the source of her family's collective illnesses in 2002. She left home for five days, and her vision cleared and other symptoms subsided. Again in 2007, her ailments abated when she left the house for three days. She then ruled out other household sources of illness, including domestic radon exposure, carbon dioxide, sewer gas, mold, and water contamination.[20] Her last-ditch attempt to ascertain the etiology of her family's illnesses was to invite a friend of a friend, named Nancy Shoemaker, who suffered from multiple chemical sensitivity. Harriett hoped that Nancy would use her heightened chemical susceptibility to divine the specific source of their health issues within the home.

Nancy, who spoke with a delicate and slightly nervous poise, had developed chemical sensitivity at an early age while attending beauty school in Nebraska. Nancy would lose consciousness and collapse nearly every morning as she sterilized the styling utensils. She had to drop out and readjust her dream of becoming a beautician. Nancy did not think much of her fainting spells until years later when she moved to Florida, where she and her husband took up residency in a trailer. After moving into the trailer, her sensitivities dramatically escalated—but not just at home. A whiff of cologne on the street or the act of shaking hands with someone wearing a transparent Band-Aid could wilt Nancy to the ground. Her body became jarringly attuned to the vast chemical infusion of the world around her.

As a result of these continual chemical encounters, she learned to move through the world with caution. When barefoot at home, she would cross sections of linoleum with great care, unsure of the daily caprice of her sensitivities. Her corporeal vulnerability to chemical vapors or direct contact is not spread uniformly throughout her body. As a high-frequency exposure site, an extra-sensitive area in the center of Nancy's palm became more acutely affected over time. Nancy took advantage of the embodied insights of her palm, honing its reactivity as an index of the hazard of various materials and spaces that she encounters in daily life. As she spoke, her gaze turned down to her hands, and she ran her right index finger in circles around the area on her left hand. "If I put something on that sensitive spot or touch something with that sensitive spot, I can tell if I can handle it at that time or not."

To manage anxiety about her emergent reactivity, Nancy developed a deeper literacy of the chemical world by way of a deeper literacy of her own body. "I know about formaldehyde and I'd never done anything like [what I did] with Harriett," she explained, "but I knew how formaldehyde affected me." She amassed somatic knowledge about formaldehyde via years of enduring its effects and affects—through dozens of fainting spells, bouts of wooziness, enervating weakness, and daily somatic tests of the material things that populate her world.

Nancy and Harriett began assessing the chemical constitution of Harriett's home. They relied on the sensitive spot in Nancy's hand as an alternative to expensive and inaccessible scientific instrumentation. Sitting in her small and immaculate assisted-living apartment, Nancy recounted the process: "And so I went into the different rooms and I tested the carpet and doors. . . . I went into the kitchen, and I just grabbed hold to open the cabinet or something. I

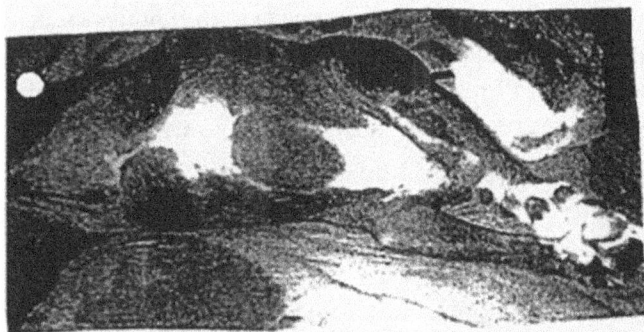

FIGURE 2.4. A photocopied entry of the records kept by Harriett McFeely, showing photos of Bowser the dog and notes. Bowser's body and disposition index the presence of otherwise-invisible chemicals.

don't think I touched it very long. . . ." At that point in the assessment, Nancy lost consciousness. Harriett observed Nancy clutch her stomach and let out a groan. The color drained from Nancy's face as she dropped to the floor and began to seize. Harriett's Boston terrier, Bowser, ran into the room to investigate the commotion and he, too, curled into a fit of seizing as he approached Nancy. The two lay there next to each other on the carpet, gripped by spasms, for a few moments before Harriett and her husband dragged Nancy outside. Bowser continued to convulse in the kitchen. The dog came to within an hour but remained disoriented, running into the furniture, walls, and doors.

Nancy gradually regained her composure over the course of half an hour. After she felt well enough, she went on her way, confident that she had found at least one source of the McFeelys' suffering. As unnerving as the experience was, Harriett also felt relieved that Nancy had validated her suspicion that chemicals were quietly emanating from her home. With an affirmative nod, Harriett emphasized the instrumentality and accuracy of Nancy's body: "In my opinion, that lady is like a human Geiger counter." Of course, Harriett, and all exposed and affected bodies, also bears this capacity to manifest the chemical world, albeit in less eventful ways. Some bodies exclaim, while others speak in hushed tones. In domestic chemical exposures, bodies are both the means of apprehension and the site of damage. Bodies uncover invisible toxicants with their wounding. Humans and their nonhuman companions serve as their own canaries in the unwitting coal mines of residential America. A month after Nancy's visit, Harriett's fifth dog in twenty years had to be put to sleep after he became wracked with near-constant seizures. The McFeelys have since lost two more dogs to similar ailments.

Like Linda, Harriett felt the pull of the late industrial sublime. She felt the attrition in her own body and monitored the bodily ailments of her dogs and her husband. In line with what the sociologist Phil Brown (1997) has called "popular epidemiology," or the grassroots appropriation of expert means of environmental health assessment (see also Murphy 2006: 62), Harriett sought to comprehend the systemic nature of such exposures. Harriett wrote letters to the editors of newspapers in five or six nearby towns. Her short notes, published in 2008, read: "Modular home owners, have you had any health problems? Have your indoor pets had any mysterious illnesses? Please write or call me." Phone calls began rolling in, one after another. Harriett began to systematically survey respondents. She asked those who called her how long they had been living in their home and what their symptoms were. She surveyed thirty individuals from thirteen different households throughout Nebraska. Respondents supplied thirty-two different symptoms they perceived to be correlated to the occupation of their modular home, ranging from unusual thirst to cancer. Harriett further inquired about indoor pet health and recorded the symptomatology of fifteen animals in seven households. She was able to garner funds for formaldehyde test kits from the Sierra Club and tested respondents' homes. Seven of the thirteen homes tested had levels of formaldehyde in excess of 81 ppb, the World Health Organization's maximum recommended exposure for half an hour. Harriett mails copies of her data, adorned with a row of skulls and crossbones along the spreadsheet's bottom border, to anyone she thinks may be able to help.

Harriett made her husband promise that he would have a thorough autopsy performed on her if she were to "drop dead" before him. Shifting her gaze over to me, she asserted that the decomposition of their dogs' bodies was a herald of her and her husband's future. "I would bet you a hundred thousand dollars that if they did an autopsy on us today, I would bet money that it is exactly like the dogs."[21] Harriett implies that their domestic exposures have reduced her and her husband to the walking dead, that a postmortem examination could rightfully be performed on them at any time. A grim suggestion, perhaps, but one that represents the outlook of many residents persevering in potentially chemically contaminated homes. As evinced by Harriett's future autopsy results, sustained chemical exposures beckon death, but they also render death ambiguous. She takes the logic of bodily reasoning to its conclusion: If wounding indicates the source of harm, then death will surely disclose its ultimate truth.

Coming to corporeally comprehend one's environment does not always have consequences as severe as in Harriett's case. Residents of potentially contaminated homes I met across the United States gradually became aware of minor departures from their normal sense of taste, sense of balance, clarity of thought, memory, durability of skin, or frequency of contracting colds. Occasionally, inhabitants claimed to have not experienced even the slightest deviation from their typical physical state, even as they recognized atmospheric irritation as an altogether-indistinct feeling. As one North Dakota man noted, "Something about the air in here doesn't seem quite right."

In these spaces where enduring and knowing are coterminous, the feeling of living death seeped into the margins of life for those with even minimal symptoms. After I spent three days in the spring of 2012 with a woman living in a resold FEMA trailer outside Oklahoma City—who was experiencing memory loss, cloudiness, fatigue, and respiratory irritation—she offhandedly mentioned that she was working on a zombie novel. We sat in lawn chairs next to her trailer, where we overlooked the highway and shared an orange. She summarized the plot: "It's not your usual blood, guts, and mayhem zombie novel. You see, after the government releases some chemicals, everyone turns into a zombie." She laughed, "They say, 'Write what you know.'"[22]

Sublime Limits

To somatically apprehend formaldehyde exposure means to begin apprehending the costs of late industrial infrastructures, economies, and standards of living. It sets in motion an appreciation that the molecular cohabitants that physically hold our world together also encourage our unraveling. Focusing on slight sensations and dysfunctions reorients chemical phenomenology toward an apprehension of the irritating chemical background noise of everyday life.

Ambient formaldehyde makes itself known to mammalian life through minor effects and affects that the exposed can accumulate, over repeated incidents, into bodily attunement to the scale of chemical saturation, beyond the individual pocket of air we call home. This string of intimate sensations, coalescing into a late industrial sublime, can inspire attempts to live differently.[23] The late industrial sublime does not merely refigure a form of the sublime from Enlightenment-era philosophical discourse but poses an alternative schema of eventfulness or call to action, one that expands dominant ideas of catastrophe and the disturbing. It opens the door to apprehending homesickness and putting into motion plans for survival, repair, and change that are charted in the rest of this book. This chemical instance is just one of many emergent approaches of reckoning with the temporally and spatially dispersed violences of contemporary political orders, which include climate change, biodiversity loss, extractive labor practices, and social abandonment, among others (Povinelli 2011; Dempsey 2013; Wells et al. 2021; Bond 2022; Zee 2022; Davies et al. 2023; Shadaan 2023).

Yet, with formaldehyde production and consumption infrastructure largely locked in, and without the capacity for networking the atomized populations charged by this form of the sublime, it remains difficult to imagine how we might, at a societal level, decamp from spaces conditioned by uncountable formaldehyde microemissions. Doing so requires massive reconceptualizations of late industrial life and the values that guide intra- and extrahuman relationships (see chapter 6). Beyond instrumentalizing viscera to overcome fetishized desire for exposure—overcoming our attraction to a new-car smell, for instance— such attunements to encounters between airs and bodies constitute the openings through which to grapple with the composition of our world and with the untold caustic ecologies that remain largely insensible to humans.

However, under dominant regimes of regulating and mitigating exposure, such pleas are either actively disqualified, as was Linda's experience, or they passively languish without gaining authority, as has been Harriett's experience. To understand the potential for sublime articulations of distributed violence to interdict ongoing harms and compel different approaches to building home, we must understand the countervailing currents of un-knowing exposures. The emergent tributaries of harm apprehension are met by highly patterned and effective practices—stemming from industry and government—for shutting down avenues for action. These anti-epistemological practices, which I discuss in the next chapter, are the opposite of the practices of bodily attunement outlined in this chapter. The bureaucratic and scientific production of surreal data sanctions the toxic home as safe and aims to quell the revolutionary potential of the attuned body and the late industrial sublime.

THREE

Un-Knowing Exposure

Paul Stewart moved his grandmother-in-law's china to the bathroom under the stairs before evacuating to higher ground. He assumed that losing the roof to high wind would be the most likely damage from the incoming hurricane. After Katrina pulverized the coast and then dissipated into a constellation of smaller storms and tornadoes, he and his wife, both white, returned to what was left of their home in Bay St. Louis, Mississippi. The second floor stood in delicate suspension over a largely barren foundation. The storm surge had erased the first story, while the faint lines of a freshly vacuumed carpet lay wholly intact upstairs. The tin roof that Paul was sure would be the first to fail was the only thing holding the remains of the house together.

Their Federal Emergency Management Agency (FEMA) trailer, delivered three months after the storm, greeted him and his wife, Melody, with the customary "strong new smell." This despite the fact that they had a friend air out the trailer for two weeks. Melody kept waking up with a runny nose their first night in the trailer. Eventually, she turned on the light to find that it was blood, and not mucus, seeping from her face onto the pillow. These small hemorrhages continued for the duration of her stay in the trailer. Paul also began to submerge into illness, with burning eyes, a scratchy throat, and an incessant unproductive cough. But too much was in flux, and too much work needed to get done, for the two of them to spend time sleuthing out the cause of their ongoing corrosion. Every corner of their lives, strewn awry by the storm, harbored a potential explanation for these inchoate caustic happenings. Did they stem from eating meals prepared by the Red Cross three times a day? Stress? The dryness of no rain for months after the storm? Contamination from the DuPont facility directly across the bay? Irritation from cutting apart trees and other debris strewn across their yard for eight to ten hours every day? How could they even begin to narrow down the etiology?

About two weeks into their stay, they awoke to an unusual silence. Dawn was usually announced by the shrill squawks of their pet cockatiel, who had long ago assumed the role of alarm clock. Upon inspection, Paul found the bird off its perch, lying on the floor of the cage. Its tongue hung out and a puslike fluid oozed out of its beak. Over the phone, the veterinarian told him to get the bird out of the trailer: The culprit might be formaldehyde. Within hours of being taken outside, the bird began to emerge from its end-of-life stupor. Years later, as he was walking through a small-town harbor in New England near where he was finishing law school, Paul recalled that as the cockatiel slowly began to regain consciousness, he began to wonder "if this is happening to him, what's happening to us?"

Trailer manufacturers won over $2 billion in no-bid contracts to supply temporary housing units to FEMA. They produced homes at a breakneck pace. A single plant owned by Gulf Stream, Inc., the company that secured the largest bundle of federal housing contracts (over $500 million), produced a hundred homes a day. Line managers marched the floor with bullhorns, yelling at workers to work faster. Laborers found it difficult to breathe on the assembly line. Some bled from their noses; others, from their ears. The fiberboard and Luan paneling were so thick with formaldehyde-based resin that it not only stunk but was also sticky and physically difficult to separate. Some fifteen years later, a woman who worked for the company in Elkhart, Indiana, that sold laminate paneling to FEMA trailer manufacturers tracked me down to confess that she had been ordered to mislabel engineered woods as "interior grade," implying that they were low-formaldehyde emitting, when they were not. The demand of massive federal contracts had blown through their existing stock of lower-formaldehyde interior-grade plywood. Requisitioning, producing, and barge-shipping new panels from their supplier in Indonesia would have taken six months. Instead of forgoing the windfall profits of disaster capitalism (Klein 2007), they decided to lie, and not only to manufacturers of FEMA trailers. From their base in Elkhart, the trailer manufacturing capital of the country, they lied to everyone who purchased their paneling in late 2005 or 2006. The toxic impacts of disaster response fell not just on those whose lives were upended by the hurricanes, but also on many others who simply chose to buy a trailer at an ill-fated time.

As government checks streamed into manufacturers' coffers, FEMA trailer inhabitants were gasping for air and information. In March 2006, a FEMA trailer resident in Jefferson, Louisiana, wrote, "It burns my eyes and I am getting headaches every day. I have tried many things, but nothing seems to work. PLEASE, PLEASE HELP ME!!" as a comment on Gulf Stream's website. The message was forwarded up the ranks to the company's president, Daniel G. Shea, whose salary doubled to over $1 million per year in 2005, where it remained in 2006. By this time, one of those sufferers, Paul Stewart, had found a low-cost means to test for formaldehyde in the form of a small badge used to assess formaldehyde exposure in hospitals and funeral parlors. The result from the Stewarts' Gulf Stream FEMA trailer was 220 parts per billion (ppb), more than double the US Environmental Protection Agency (EPA) maximum recommended threshold, which was itself twelve times the minimum response level for chronic exposure (see appendix 1).

Emboldened with this new data, Paul called FEMA. The agency held firm in its position that the trailers were safe and suggested that perhaps Paul was chemically sensitive. He called an acquaintance who worked for a Republican congressional representative, who, after speaking with FEMA, relayed to Paul that "FEMA is not concerned because they have sovereign immunity." On March 16, 2006, Paul took his story to WLOX, a television news station in Biloxi, Mississippi. Later that day Shea, Gulf Stream's president, wrote a surreal missive to one of his contacts at FEMA, stating, "Gulf Stream is one of the leaders in reducing formaldehyde emission in recreational vehicles."[1] The following day, as manufacturers began hiring public relations firms and crafting statements of denial, Paul received a call from a local environmental activist, Becky Gillette, also white, who knew him from environmental organizing and had seen him on TV. Becky was direct: "I can't believe what I just saw. I think we should test them."

Becky recounted these events to me in the spring of 2011 in the small Arkansas town that she now calls home, her feet swathed in thick socks with peace signs at the ankle.[2] Together, Becky and Paul secured funding from the Sierra Club and used the same lower-cost test kits that Paul had found online to measure the formaldehyde in sixty-nine FEMA trailers in Mississippi.[3] They found that 88 percent contained indoor formaldehyde in concentrations higher than the EPA's 100-ppb threshold. This grassroots, community-initiated assessment drew international media attention and incited a crisis in postdisaster governance and a multimillion-dollar lawsuit. By May 16, the story had become a major feature of CNN's post-Katrina coverage. Two days later, the first lawsuit related to the FEMA trailer formaldehyde exposure was filed (*Hillard v. United States Government*, Complaint, Civil Action 06-2576). Eventually some ninety thousand people would join the ranks of litigation.

As this knowledge relay from bodily knowledge to community science to media and legal attention progressed, it encountered interference from the domain of society that is most squarely charged with determining the reality and impacts of these exposures: science. Far from being a clean tool of logic and rationality, this chapter details the often-surreal ways that the government and industry deployed science to un-know what bodies and low-cost sensors had indicated to be true. Specifically, I track the multiple scientific processes that refute, dilute, disqualify, or simply avoid claims of FEMA trailer–related formaldehyde exposure and ensuing negative health outcomes. Multiple and divergent technical practices weave together into a larger regime of chemical exposure and ensuing illness obfuscation that I will refer to as *un-knowing*. The means of un-knowing that I document span both federal postdisaster governance and the application of science in court. While the actors, instruments, and techniques of un-knowing shift between the initial government response and the courts, the outcomes remain the same. The analysis brings into further relief the negative epistemological space that shapes the contours of both technoscientific and legal facts.

This manifold character of un-knowing is part of its strength. Un-knowing is not always and everywhere about refuting and avoiding knowledge; rather, it also involves assembling superseding knowledge claims that gain their force from cultivating authority rather than empiricism. Un-knowing, in other words, involves both intentional tactical ignorance and discursive ignorance. When combined, this string of reactive tactics and opportunisms is extremely effective.[4]

Forestalling Assessment

In May 2006, a FEMA spokesperson placidly announced, "FEMA and industry experts are monitoring the small number of cases where odors of formaldehyde have been reported, and we are confident that there is no ongoing risk" (Associated Press 2006). Behind the serenity of their official statements, FEMA was divided in planning its response. Since mid-March, internal emails circulating among FEMA field offices advised an immediate and proactive response. When FEMA trial attorney Patrick "Rick" Edward Preston was assigned to the case in June, he was propelled to the forefront of FEMA's scientific inquiry into the formaldehyde issue (Minority Staff Report 2008: 10).

One day after he was appointed to the litigation team, Preston sent an email vetoing swift evaluation: "Do not initiate testing until we give the OK. While I agree that we should conduct testing, we should not do so until we are fully prepared to respond to the results. Once you get results and should they indicate some problem, the clock is ticking on our duty to respond to them."[5] The institutional liability of verifying the hazards of chemical exposure outweighed such knowledge's utility in informing citizens or minimizing the corporeal risks of exposure that continued to accrue with time. This form of forestalling the scientific and systematic corroboration of FEMA trailer-related illnesses was the first and most straightforward instance of the agency's un-knowing of exposure.

In June 2006, FEMA, the Agency for Toxic Substances and Disease Registry (ATSDR), the EPA, and the Centers for Disease Control and Prevention (CDC) began conducting biweekly interagency conference calls to address escalating public relations, public health, and litigation concerns about formaldehyde exposure.[6] Through these calls, it was decided in August that the EPA would run tests in September and October (a full year after the landfall of Katrina) of *unoccupied* trailers to test what methods of ventilation were most efficacious—a study design that, as ATSDR officials told me, was the result of directives from FEMA attorneys. The formaldehyde litigation was a central catalyst and concern of the study, as indicated by handwritten notes from the calls by one of the ATSDR officials that coauthored their report.[7] The politics of the study design were clear: the assessment was to avoid the health risks posed by the actual inhabitation of the trailers in pursuit of gauging ways that the trailers could hypothetically be made safe.

At the time, Gillette, as a central organizer of the Sierra Club's testing efforts, felt triumphant that the group's efforts had garnered federal attention: "I was happy when I heard the EPA was testing in October. I was glad that the professionals were coming in to verify." Her faith began to ebb as the new year arrived but no results were made public. She explained, "They really dragged their feet and so I wrote a FOIA [Freedom of Information Act letter] in February to force them to release the results." The EPA had conducted the tests, but the agency had then delivered the data unanalyzed to FEMA, which was then supposed to forward them to ATSDR for analysis. In early December 2006, ATSDR was still awaiting delivery of the test results from Preston's office.

Soon thereafter, Preston passed on the data. But before they forwarded the monitoring data, FEMA strictly stipulated that all ATSDR analyses would remain confidential. "No information should be released to any third party without my express permission," Preston mandated in an email to ATSDR.[8] The report remained duly secret until FEMA issued a press release in May 2007— three months after ATSDR sent its final report to Preston's office and eight months after the tests were conducted.

Leveling Concern

Setting the correct hazardous exposure threshold for FEMA trailers had been contentious from the start. When ATSDR analyzed the potential hazards posed by the indoor air quality of unoccupied FEMA trailers in 2007, they disregarded their own intermediate (14–364 days) and long-term (> 1 year) formaldehyde minimum response levels of 30 ppb and 8 ppb, respectively. Instead, in drafting the report, the authors crafted their own standard, which they dubbed "the level of concern" (ATSDR 2007). The level of concern was set at 300 ppb—ten to thirty-seven times the concentration of formaldehyde that the agency had deemed capable of causing adverse health effects. This threshold is also three times higher than the level that the EPA, National Cancer Institute, and the Occupational Safety and Health Administration assert can cause not just irritation but coughing, skin rashes, and severe allergic reactions.

According to former ATSDR Director Howard Frumkin, this level of concern bore "little or no operational meaning" (Minority Staff Report 2008: 17). Even generally accepted thresholds of significance harbor enough uncertainties to beg the question, "For *whose* standards, and by what version of proof is a 'standard of proof' determined or employed?" (Brown 1992: 275, emphasis in original), but this ad hoc threshold would appear to be an even more urgent site of condensed politics. Setting the level of concern at 300 ppb dismissed the illnesses that stricter guidelines were intended to prevent, labeling them mere "nuisance" symptoms. ATSDR's report found opening all windows, static vents, and exhaust fan vents would, on average, reduce the indoor formaldehyde levels of FEMA trailers to below their "level of concern." Hence, the trailers should not be expected to produce adverse health effects (ATSDR 2007).

This tidy conclusion served to temporarily quell rising concern about the chemical consequences of inhabiting a FEMA trailer. Notice that the study was *not* designed to assess the chemical hazards of FEMA trailers as typically inhabited by humans, but to see whether it was theoretically possible to bring the indoor formaldehyde levels within a "safe" range. The agency based its conclusion of safety on a set of test-case conditions that were unlikely to occur either during brutal Gulf Coast summers, when residents close their trailer windows and run their air-conditioning at full tilt to fend off withering heat and humidity, or in winter, when residents must close their windows to keep warm.

Frank Renda, then a senior environmental scientist at ATSDR and leader of its Division of Toxicology and Environmental Medicine, was alarmed by the report when it landed on his desk for approval: "[The consultation] was dated February 1st [2007], I saw it on about the 17th of February; within about two hours I called my senior management and said that we had a problem. The consultation as it was written was misleading, possibly misleading, and a potential public health threat." Renda zeroed in on the report's silence on long-term health risks, including cancer, reproductive development, or sensitization (that is, how a short-term exposure can make someone later experience health effects at much lower doses).[9] As Renda recalled over lunch in Atlanta in 2011, he traced the study design's lineage to the FEMA attorney, Rick Preston, who requested the ATSDR analysis: "They had been directed by FEMA first of all not to share it with anyone [and secondly] that they were only to address the shorter term. That was the thing; I didn't have to go into any in-depth review to know that we had missed the boat, we missed the mark, that FEMA had gotten what they had asked for." Neither Renda's technical misgivings nor the patent influence of a lawyer managing FEMA's liability moved Renda's superiors to amend the consultation. The report was sent to FEMA and only resurfaced later in the summer when, as Renda told me, "things began to break loose."

In May, FEMA invoked the report in a press release justifying inaction.[10] In response to a critical reception, ATSDR's senior management raked the two junior scientists who had penned the report over the coals (they would later be promoted). Hoping to silence criticism that their agency had chosen to protect the government and its corporate vendors over the people they were charged with protecting, ATSDR's senior leadership then asked Renda to lead a cross-agency working group to develop recommendations for remediating the report.

Renda's committee produced a revised document that bore increased precautions. As he recollected in between bites of salad: "One of the points stated, 'Given the hazards posed by the trailers and the formaldehyde exposures, that efforts should be undertaken in the areas of health education and that appropriate measures to interdict exposures should be implemented.' And after that there was, all of a sudden, some sudden displeasure with what I had done. It was removed from my oversight and the executive summary was revised to say 'analyze' as opposed to 'implement.' Paralysis by analysis." Renda was removed from his role supervising the revision of the ATSDR report on the FEMA trailers. Ten days after he wrote this memo, Renda was notified that he had been given an unsatisfactory professional evaluation and removed from his position. He recounted the sequence of events as follows:

> I was on my way to Italy. I am a member of something called the Collegium Ramazzini [an elite academy of 180 environmental and occupational health scientists] and I was there with my father. His parents had come as immigrants from Italy around the 1900s and he had never been [to Italy] and I get to take a guest so I said, "Why don't you come as my guest?" We were at the final awards ceremony and I was approached by one of my managers with a double blue folder: One side is my unsatisfactory evaluation and the other side is my notice of removal. My father is there, my colleagues are there, seeing all of this. So that was complicated.

I asked if being reassigned in such a public and poignant manner was punitive. Renda took a deep breath and, with his right index finger, began tracing the outline of the metal links on his watch. "I think it could be construed to be mean-spirited. But, uh, I can't say that with certainty. So that would be a construct that others would have to. . . ." He trailed off as his eyes welled up behind his metal-rimmed glasses as he held back his opinion. One of the conditions of his termination settlement was that he was to say nothing negative about the agency.

After he was removed from his position at ATSDR, Renda was placed on "a performance improvement plan"—a slow, bureaucratic means of firing. As he waited for his motion of wrongful removal to be processed through the court, he was repeatedly reassigned to different offices within ATSDR's sister agency, the CDC. "I ended up in an office in a new building with limited ventilation," he recalled. "I complained about the air quality and it turned out by their own measurements by *their* hygienist, the levels of formaldehyde were twice the recommended limit by the National Institute of Occupational Safety. . . . I was developing rashes [of] which I still have some." In a back office of the CDC's National Center for Environmental Health, Renda was slowly infused with the chemical that cost him his job. The substance he was reprimanded for simply suggesting taking precaution against seeped back into his life, more materially than ever, leaving its mark on his career and his corpus. He was struggling to right his toppled professional life when I last saw him in the spring of 2011.[11] Perpetual analysis and the suppression of not just dissident views but the people that advance them were key to un-knowing formaldehyde exposure.[12]

Further criticism was leveled against the 300-ppb "level of concern" by Dr. Vincent Garry, a pathologist who had previously reviewed ATSDR's 1999 toxicological profile on formaldehyde. In a March 2008 letter to Representative Brad Miller, the chairman of the Subcommittee on Investigations and Oversight, Garry pointed out the inappropriateness of using a threshold developed for occupational safety to establish the safety of indoor air in residential units. Occupational exposure studies are based on eight hours of exposure per day for five days a week, whereas, for many trailer residents, "this is a 24 hr per day 7 days per week exposure" (Garry 2008: 1). Garry also noted the physiological differences between the largely healthy adult males who composed the occupational studies and the wide-ranging biological makeup of trailer inhabitants, writing, "For example, children under age two have a short trachea and breathe faster than adults (thirty to forty breaths per minute), therefore, process more formaldehyde into the body and are probably less efficient in the metabolism of the chemical" (1). His letter concluded by blaming these errors on the lack of both peer review and robust intra-agency communication.

The issues that Garry and Renda highlighted are not isolated to this particular incident. Rather, they are symptoms of a long-standing lack of oversight and shortcomings in the design of chemical exposure studies. From its early days in the 1980s to the present, ATSDR has been accumulating complaints regarding its failure to meaningfully investigate community concerns about toxic exposures (ATSDR Staff Report 2009). The US Government Accountability Office (GAO), in a 1991 review, found ATSDR's results to be "seriously deficient as public health analyses" (GAO 1991: 2). The Environmental Health Network and the National Toxics Campaign Fund asserted in a 1992 study that ATSDR's studies were "inconclusive by design," supporting their argument with multiple accounts of toxic exposures in the Gulf Coast that were deemed by ATSDR to pose no health risk (Russell et al. 1992). Corroborating this statement, one ATSDR scientist, who bore witness anonymously, testified that "it seems like the goal is to disprove the communities' concerns rather than actually trying to prove exposures" (Subcommittee on Investigations and Oversight 2009: 2). From asbestos contamination on a beach in Chicago, to depleted uranium in Upstate New York, and the emissions from cement kilns in central Texas, the federal investigations that communities worked so hard to initiate left exposed residents distrustful of the agency's ability to investigate their concerns in good faith.

The EPA data, collected in trailers that were ventilated to an extent beyond practical possibility, was an apparatus of hazardous knowledge avoidance. The threshold and the policy decisions it justified subjected tens, if not hundreds, of thousands of Gulf Coast residents to sustained formaldehyde exposures, for "at least one year longer than necessary" (Minority Staff Report 2008: 1). The level of concern further dismissed the existing scientific recommendations for maximum domestic formaldehyde concentrations—including those of ATSDR itself—as overly cautious and attempted to mask the hazards manifesting in the exposure data it was charged with adjudicating. These technical maneuvers, moreover, must be understood within ATSDR's distinct history of minimizing the public's concerns about cohabitating with toxics in favor of reducing the liability of fellow federal agencies, of prioritizing the needs of industry over public health.

The day before a congressional hearing on "FEMA's toxic trailers" was slated to begin in mid-July 2007, nearly two years after the storm and two months after the "level of concern" ATSDR report was released, FEMA admitted that it had made a mistake. In an effort that many commentators interpreted as damage control to protect FEMA's director at the hearing, FEMA announced that it would provide funds for the CDC to test occupied FEMA trailers. From late December of that year to January 2008—the months when ambient temperatures and therefore formaldehyde off-gassing levels are lowest—the CDC tested 519 FEMA trailers. These tests confirmed elevated formaldehyde levels. The geometric mean of these tests was 77 ppb, just shy of the WHO's maximum recommended exposure of 81 ppb for half an hour (CDC 2010). The findings underrepresented both the past, as off-gassing slowly decreases in the first eight months after manufacture, and the future, as summers increase exposure with higher temperatures and humidity. FEMA successfully deferred testing trailers in the conditions of inhabitation until the last possible moment in order to not endanger the careers of its leadership.[13]

Arguing Asthma

Un-knowing stems not only from the often-blunt ways that state entities employ liability avoidance, deferral, or disqualification as techniques of governance, but also from the more indirect ways that both governments and private industry use scientific ambiguity and inconclusivity. A primary figure in these relatively more subtle avenues of un-knowing is the ever-shifting expertise of the "expert witness." The way science is used to produce truth in US courtrooms is often unrecognizable to professional scientists. Different interpretations of evidence are core to scientific debate. In the courts, the friction between competing scientific explanations is thrown into overdrive to the point that some have labeled it "terminally adversarial" (Albright 2023). The expert witness that leverages science in service of their client's legal desires is where the rubber hits the road for exposure science. I document this contested importation of science into the house of law through an exemplary test case in the FEMA formaldehyde lawsuit, technically known as a "bellwether trial." The trial pivoted around the question of whether formaldehyde exposure in a FEMA trailer in New Orleans exacerbated the plaintiff's asthma.

Christopher Cooper was eight years old when Katrina hit and the flood protection system of New Orleans failed, destroying his family's home. Christopher grew up in New Orleans East with his mother, Alana, and older sister, Erika, all of whom are Black. Up until the levees broke in 2005, they lived in a house built by his maternal grandfather. His grandparents ran the barbershop directly next door to their family home.

Christopher was diagnosed with asthma when he was three. In a case lodged against trailer manufacturers and installers, Christopher's family alleged that formaldehyde levels in FEMA-supplied trailers exacerbated the boy's asthma. Later, the court selected Christopher's case as a lead case—the "bellwether case"—for a larger class-action suit for plaintiffs with similar ailments. I worked on a subsequent case that dealt with adult asthma causation, oral tumor growth, and depression that presented similar epistemological techniques in the courtroom. I have nonetheless selected this case, out of many other possible cases I might have discussed, for several reasons. First, exacerbation speaks more to the question of making an existing situation worse rather than causing a new discrete disease to emerge. This form of immiseration is more closely allied to the larger questions of this book about how to grapple with harm in an imperfect, complex world of exposure where pathophysiologies already exist and accountability is never assuming a pristine baseline for comparison. Second, children are more susceptible than adults to harms related to chemical exposure but are less represented in the scientific literature compared to adults. These exposures are potentially the highest stakes and yet the most scientifically uncharted. Finally, I find it useful to discuss a childhood case because courtroom discussions of cases involving children are largely free of the murky, and often racist, science of forensic psychology and manipulated personal histories or comorbidities that all too frequently appear in adult cases.

Since his diagnosis at a young age, Christopher used his corticosteroid "rescue" inhaler once or twice a month when he found himself gasping for breath. Typically, this occurred during sports-induced asthma attacks. Christopher explained this in a deposition:

Q. How did the asthma affect what you did on a daily basis before the hurricane?

A. Well, when I go outside and play with Ms. Donna's children [the children of his mother's best friend], I will sometimes get a little wheezy and I have to maybe stop and just wait for a while so I can go back and play.

Q. When you got wheezy, did you have to use your inhaler?

A. Yes, or I might just want to stay inside. [. . .]

Q. How did you feel when you were having an asthma attack?

A. I would kind of feel tired and, like, I would just, like, want to go in my bed and go to sleep.

Q. Did the asthma attack make it hard to breathe?

A. Yes.

Q. Did that scare you?

A. Yes, it did.

Q. When you had asthma attacks before the hurricane, did you have any concerns about how it was going to affect your health?

A. No.

Q. So what scared you about the asthma attack?

A. That I might die. (New Orleans, June 29, 2009)

As the seasons changed, Christopher's asthma would "flare up." He averaged two asthma-related hospitalizations a year: one on the way into winter and one on the way out. After the hurricane displaced his family to Florida for six months, Christopher, his mother, and his older sister moved into a FEMA trailer parked in the driveway of their flooded home. Soon thereafter, the frequency of Chris's asthma attacks increased fourfold. This worsening of his asthma was the basis for Chris's legal claim.

The counsel for the defense hired Dr. Kenneth Smith, director of the intensive care unit at East Jefferson General Hospital, just west of New Orleans, and a practicing respiratory disease specialist with Jefferson Pulmonary Associates. The defense counsel hired this white doctor to refute both the general claim that formaldehyde caused asthma (*can* x cause y?) and the specific claim that formaldehyde exposure had exacerbated Christopher Cooper's asthma (*did* this x cause this y?). In the following excerpt from a deposition, one of the plaintiff's attorneys asked Smith whether a series of statements were true or false—a line of questioning often used to establish the baseline opinions of expert witnesses:

Q. First statement: There's a cause-and-effect relation between formaldehyde and asthma.

A. That's a qualified maybe.

Q. Is there a cause-and-effect relationship between formaldehyde and initiation of asthma?

A. Possibly.

Q. Formaldehyde is an irritant to the respiratory system?

A. Possibly. Again, these are all concentration issues. They are potential irritants at the appropriate concentration. We've seen several patients who lived temporarily in FEMA trailers that have asthma and have other illnesses. But that [allergen and toxin exposure] was a very common occurrence in this city after Katrina. [. . .]

Q. Formaldehyde exposure can have harmful effects on developing lungs in children?

A. I would say that I'm unaware of any data that would suggest that that is the case.

Q. So is that a no?

A. That's a qualified no because I'm unaware of it. (Metairie, Louisiana, July 10, 2009)

This strained back-and-forth continued for several minutes. To questions revolving around whether the scientific literature demonstrated that low-dose formaldehyde exposure can exacerbate asthma, Smith responded "maybe," "possibly," and "I'm unaware of it." He also, and to a lesser extent, answered with "false."

The latter was his ruling on the statement "There is a correlation between formaldehyde exposure and the risk of allergic sensitization." Allergic sensitization is one of three pathophysiological links between formaldehyde exposure and asthma attacks. The connection between formaldehyde and allergic sensitization is well established and uncontroversial (Wantke et al. 1996; Garrett et al. 1999). After Smith's firm negative answer, the questioning attorney pressed further:

Q. A definitive not true or you haven't seen any studies that suggest that?

A. I haven't seen any studies that suggest that, and the studies that I have seen suggest that that's not the case.

In Smith's list of reliance materials, a technical term for everything that he has consulted in formulating his opinion, he listed only seven articles. Of these articles, none specifically scrutinized allergic sensitization in the presence of formaldehyde, and most had little to no bearing on the trial's claimed concurrence of FEMA trailer residency and childhood asthma exacerbation. Smith produced a firm account of causal uncertainty through his listless engagement with the scientific literature, a technique of presenting political matters as technical matters that sociologists of science and law have dubbed "performing closure" (Lynch et al. 2008: 228).

Smith's window on the scientific literature was not just small but also strategically sculpted, as evidenced through his responses to questions from the plaintiff's attorney:

Q. The studies that you reviewed that are in your reliance file, how did you get those? Did you go out and hit the books or hit the computer or did the lawyers provide them? Did someone else provide them to you?

A. Most of the—the documents that I referenced in my opinion paper, except the ones for Up-To-Date, which is an on-line textbook, mega textbook—

Q. Sure.

A.—were provided by defense counsel.

Thus, Smith's intentional ignorance was shaped by the tandem efforts of defense attorneys' strategically sanitized reading list and his own restrained research. By way of this combined effort, Smith was able to achieve plausible deniability. He was able to aver, without fear of perjury, formaldehyde's noncausation of asthma. Through this calibrated epistemic avoidance, his testimony established the nonexistence of a relationship between formaldehyde and asthma as ontologically fixed.

Smith further buttressed his claims by demarcating his expertise: "I'm not a study wonk," he remarked as one of the plaintiff's attorneys questioned him about exactly which study refuted sensitization to formaldehyde. When an attorney pressed him on the specifics of a study, Smith responded, "That's not my deal. I'm a clinician." Smith has treated at least fifty thousand patients with respiratory ailments since he finished his training in 1978. In court, he centered his expertise on his capabilities and knowledge in practice, not on his methods for exploring the literature or his ability to remember specifics of a text. And yet, despite his assertions of being a hands-on expert, he reported that not a single one of his patients claimed formaldehyde-induced asthma aggravation.[14]

Later, Smith changed course, stating that he does not rely on his clinical instinct in formulating his professional opinions. When asked how he formulates his opinions, Smith circled back to the literature, which is in conflict with his clinical common sense:

Well, it's based on my personal experience, but it's also based on literature and people that have done this. You know, quite frankly, quite frankly, I—I would have thought intuitively, intuitively that any irritant can—can exacerbate asthma. All right? If you ask me as a clinician, I'd say, yeah, virtually any irritant. So it was real interesting to me as a—as a scientist—to read literature that's from good places that seem to be good literature where that in fact is not the case. . . . That's somewhat surprising, quite frankly. But I believe it.

Smith did not base his expert opinion on the nonexistence of formaldehyde-induced asthma exacerbation on what he saw every day as a clinician, nor did he claim to base his views on the literature alone.

Smith's expertise arose from a repeated oscillation between the clinical and the scholarly, yet, as he jockeyed back and forth, he persistently located the seat of his knowledge in his *other* medical existence.[15] The choreography of Smith's expertise sufficed to maintain his legitimacy in court—the presiding judge did not call it into question—and lent his knowledge a deeply surreal quality. His expertise came from neither the scientific literature nor clinical experience, yet it was also from both.[16]

The plaintiff's attorneys hired their own expert witness: Patricia Williams, PhD, then director of the Occupational Toxicology Outreach Program at the University of New Orleans. While Smith denied any, or at least very little, causal relationship between formaldehyde exposure and asthma, Williams, also white, enacted the opposite perspective at a molecular register. Williams, a toxicologist, justified moving her expertise from the population level to the chemical level by citing a "vacuum" of data on long-term formaldehyde inhalation at low doses. This data dearth is especially acute when it comes to nonnormative groups, including the immunocompromised, the elderly, and children. Williams asserted that these epidemiological studies have not been done, and also that they should not be done, given ethical concerns: The kind of scientific studies that would concretely demonstrate these links would involve provoking asthma attacks in sensitive populations. The ethical limits of experimental science underlie the knowledge gap between what is mechanistically likely and what can be empirically proven.

In her deposition, as the defense lawyers questioned Williams about the lack of experimental data linking asthma attacks to formaldehyde, Williams circled back to other ways of knowing. She listed the three causal pathways of formaldehyde-induced asthma as evidence enough:

Q. As we sit here today, experimentally it has not been proven that formaldehyde causes asthma, has it?

A. Yes, it has.

Q. In what human study?

A. Asthma is asthma. Asthma is bronchoconstriction. You have a host—we know it certainly acts with the trigeminal-vagal reflex. We know that. We know that it increases the nitric oxide. When you say asthma, you are not saying which of those three mechanisms is, in effect, IgE or trigeminal-vagal or the nitrogen oxide synthase mechanisms. We know—there are many studies, we know that it [formaldehyde] can cause asthma. (New Orleans, December 3, 2009)[17]

Asthma is asthma. Asthma is bronchoconstriction. For Williams, studies that link the mechanisms of asthma attack to the presence of formaldehyde are equivalent to experimental evidence deduced by exposing human subjects in the lab. The molecular possibility of asthma, for her, is tantamount to asthma itself. Asthma, in Williams's testimony, is both a singular pathophysiology, bronchoconstriction, and—in the case of formaldehyde—a triad of mechanisms. She disassembles the definition of asthma to the lowest common denominator, bronchoconstriction. Through this technique of rendering what asthma *is*, of managing the polyvalence of formaldehyde-asthma causal pathways and the heterogeneity of the disease's ontology, Williams articulates a basis for her assertion of general causation. Whereas Williams styles the polyvalence of asthma causation to strengthen her testimony, Smith styles the polyvalence of his expertness to justify his evasion of the literature.

Pharmaceutical Logics

In May 2009, Christopher Cooper and his mother, Alana Alexander, traveled to National Jewish Hospital in Colorado on a trip paid for by the plaintiff steering committee. The respiratory wing of this hospital is widely regarded as a global leader in pulmonary care. There, Dr. Karin A. Pacheco ran a battery of tests on Christopher: pulmonary function tests, a challenge with a bronchoconstrictive agent, skin-prick allergy testing, and CT scans of his sinuses and chest. In her affidavit, she concluded that it is "medically plausible that [Christopher's] asthma would have been aggravated by occupancy of a water-damaged trailer with formaldehyde off-gassing" (Pacheco 2009: 7). However, given the lack of baseline medical records (many of which were lost to the 2005 floodwaters), she could not attest to the precise degree of aggravation. "Nonetheless," she continues, "the patient's test results document moderately severe bronchial hyperresponsiveness as well as a component of fixed airflow obstruction. This likely reflects a long history of asthma that has been suboptimally treated. Although it is common to wish to minimize medication use, especially in children, in those with asthma this is not necessarily a good policy. Chronic untreated asthma can lead to airways remodeling and scarring, with fixed airflow obstruction that no longer completely reverses with inhaled bronchodilator" (7). Pacheco reads Christopher's pulmonary landscape as sculpted by a long-standing lack of pharmaceutical intervention. This explanation of Christopher's current airway topography falls in line with an even longer history of the biomedical imagination's use of asthma pharmaceuticals to remove environmental factors from equations of asthma causation or exacerbation, which, in this instance, essentially blames his single mother for the severity of his case.

In his wide-ranging history of asthma and allergies in the United States, Gregg Mitman notes that, following the conclusion of World War II, "engineering the body in ways that would overcome the peculiarities of place and environmental change became a consumer and corporate dream" (2007: 211). In other words, in the eyes of biomedicine, bronchodilating inhalers replaced considerations of environmental exposures when thinking about asthma or allergies. This shift in how pulmonologists conceptualized asthma aligned the interests of patient-consumers and the pharmaceutical industry and obscured many of the underlying environmental triggers. Ian Whitmarsh also notes,

"Increasingly in the early-twentieth-century United States, pharmaceutical intervention was used to define [asthma]" (2008: 61, drawing on Keirns 2004). The question of which forms of bronchoconstriction could be classified as asthma was directly tied to whether they could be alleviated by inhaled pharmaceuticals. In a circular fashion, the treatment of asthma has seeped into its definition, and a lack of treatment has become the cause of asthmatic damage to the lungs. The original impetus for respiratory distress and pharmaceutical innovation, the environment, has been displaced by its response. Under this pharmaceuticalized (Biehl 2007) conception of nonenvironmental causality, responsibility for Christopher's asthma falls on his mother. Pharmaceutical logics align with and advance longer histories of the raced and gendered blaming of Black women, known as misogynoir (Bailey 2021), for children's illnesses caused by structural factors (Jackson and Mannix 2004; Alexander and Currie 2017).

In court, Pacheco appeared only in an affidavit, not in person. The plaintiff's counsel brought in her expert report as an exhibit. The defense counsel then had an opportunity to ask its witness, Smith, to translate Pacheco's medical terminology into more accessible English. At the beginning of his analysis, Smith paused to "applaud" Pacheco's reading of Christopher's severe bronchial hyperresponsiveness and fixed airflow obstruction as a function of a lack of pharmaceutical intervention rather than as a pathophysiology exacerbated by environmental exposures. Indeed, Christopher's mother's responsibility for the changes to his airways is, Smith posits, evinced by "the fact that he has no alteration in forced expiratory volume in one second that doesn't go normal after the use of a bronchodilator." In other words, how much air he can exhale at a baseline is the same as how much air he can exhale after a chemically triggered bronchoconstriction is countered by a fast-acting asthma inhaler. Christopher's test responses place his bronchial sensitivity snugly within the pharmaceuticalized definition of asthma. His body should, therefore, be able to remove environmental encounters with asthmagens from asthma's causal horizon.

Contra Williams's assertion that "asthma is bronchoconstriction," Smith implies that pathophysiologies are not asthma; they are merely risk factors for asthma. Asthma, in Smith's testimony, becomes a lack of human control of bodily risks. He testified, "Christopher actually has mild asthma. He has severe bronchial hyperresponsiveness which may put him at risk for more severe asthma or may—if he's not controlled, he may—he could have long-term sequelae." Christopher's asthma is itself mild, although his lungs are very reactive. As Smith depicts it, asthma is not in the body or influenced by the body's surroundings, but rather comes into being by how one medicates or does not sufficiently medicate the body.

In his expert testimony, Smith aligned diffuse logics of pharmaceuticalization with his own argument. This marks a profound cross-industry synergy between pharmaceutical companies that make asthma inhalers, petrochemical companies that make products that could otherwise be blamed for asthma causation, and misogynoir. The strategic avoidance of knowledge, corporate redefinition of disease etiology, and larger cultural practices of ascribing Black women responsibility for structural discrimination all work toward unknowing Christopher's asthma.

Inconclusion

Un-knowing managed to scientifically disappear a form of homesickness detected by the bodies of hundreds of thousands of people. The surreal textures of exposed life extend into science itself, constraining the political horizons of the late industrial sublime. These un-knowing practices are mercurial. They take multiple forms across scale and time, including secrecy, assessment postponement, scientific disqualification, dissident subjugation, knowledge avoidance, and the medico-corporate obfuscation of environmental triggers. A focus on processes of un-knowing highlights how willful knowledge avoidance and repression and discursive blinders, such as pharmaceuticalization and discursive misogynoir, work hand in hand and toward similar ends, even without clear intentionality.

For those who lived in a FEMA trailer, "the horrible unending of not knowing" that came to characterize all of post-Katrina life (Winkler-Schmit 2006) was extended by many forms of science. As the years pass, it has become more and more certain that the health effects of massive governmental toxic exposure will remain unknown. Some three years after ATSDR performed the initial formaldehyde hazard assessment that lies at the heart of this chapter, the CDC awarded a $3.5 million contract to TRI International, a large research and technical services nonprofit, to conduct the feasibility phase of a longitudinal health study of children, like Christopher Cooper, who had lived in FEMA trailers. The study was funded by FEMA (via the CDC) and known as Children's Health After the Storms, or CHATS. The organization's final report, issued in 2013, indicates that the study met and exceeded their feasibility criteria and made recommendations on how to best implement the full study. The actual study has yet to occur. In response to my FOIA requests attempting to understand the CDC's plans for full implementation of the study, the CDC was clear that "the decision on whether to move forward with a full registry ... rests fully with FEMA."[18] In addition to secrecy, assessment postponement, scientific disqualification, dissident subjugation, knowledge avoidance, and the pharmaceutical obfuscation of environmental triggers, simply starving meaningful analysis of funding may be the final means of un-knowing the likely largest formaldehyde exposure in human history. Long before the federal government failed to fund a study of the wounding it caused, tens of thousands of exposed Gulf Coast residents turned to the courts in attempts to cut through the surreal governance of the state and adjudicate responsibility for their atmospheric immiseration.

FOUR

Environmental Litigation and the
Fantasy of Accountability

As I descended her trailer's cantilevered front stairs, the entire home shudder-
ing with each step, Tina asked me a very simple question that I couldn't answer.
With a baby on her hip and a shirtless toddler rummaging through the tan
earth by her feet, she held my eyes with an unyielding gaze as I exited her home.
Without breaking eye contact, she extended her arm toward the trailer, a former Federal Emergency Management Agency (FEMA) unit, that she and her
husband had purchased in Arizona after it wound its way through gray markets. "This is your life's work," she said, her free arm still outstretched, pointing
toward the trailer. "*Who* is responsible for this?" (fig. 4.1).

Weeks earlier she had sent me screenshots of her text-message conversation
with the woman who had sold her the trailer. The woman chided Tina, writing,
"You should have done your research better," implying that it was Tina's fault
for not knowing the infamous history of the FEMA trailer. But the oversight
had been shared, as the seller admitted that she, too, was oblivious of its history
when she bought it. She "should have been smarter to look into it." If Tina filed
a civil suit, any court-ordered refund or damages would come from the seller's
pockets, which were empty. "no $ so do ur thing," she texted Tina, daring her
to take the dispute to small claims court.

Despite Tina's deep and prolonged anger toward the seller, who was both
a swindler and had been swindled herself, she knew that the onus of account-

ability for government-issued goods must not fall only on her shoulders. Clawing back and forth for scraps of capital that neither of them had was not the remit of justice. Before tracking me down, but after her family began to fall ill, she spent eight months calling and writing to every acronymed federal agency that might have authority over this issue: HUD (Housing and Urban Development), FTC (Federal Trade Commission), FEMA, FBI (Federal Bureau of Investigation), and CDC (Centers for Disease Control and Prevention). She had called the local sheriff, legal aid, and the state attorney general. This is a common routine before people become desperate enough to find me. Tina told me that I was the only one to ever write her back or return her calls. When she wrote to me, I realized that there was one federal stone left unturned. I called the Office of the Inspector General at the General Services Administration (GSA) on her behalf, because the GSA sold the trailers on the open market, and it was their sales certifications that required that these trailers *not* be sold as housing in perpetuity and that all subsequent purchasers must be informed of the health risks of inhabiting these structures. But they didn't call me back either. They never responded to Tina's further requests.[1]

As we talked through the nearly exhausted slate of institutions that might mediate the return of her trailer, Tina's oldest son looked up from where he was playing by her feet with some small toy and asked with all the sincerity a toddler can possess, "Am I one of the good guys?" It was a refrain I'd hear him loop back to over multiple visits. Tina scooped him up to give him a reassuring squeeze and bouquet of kisses: "Of course, you're one of the good ones!" Buoyed in spirit, the alabaster toddler sought out shade. Counterintuitively, he put his shoes on to go inside: The metal treads on the trailer's steps were often inexplicably electrified. I could see how it might be hard for him to know what role he's playing when his shelter is a suspected site of exposure, when the "good guys" of the state that are charged with his protection also generated the hazard that surrounds his life, when his mother wonders out loud if life would be better if the family left the first home she had ever owned and instead became homeless again, living under a juniper bush. The yawning chasm between how he is told that accountability and security *should* operate in the hero children's TV programs he watches and the way the world around him operates could only be disorienting.

Tina received a summons for jury duty shortly before we first met. Turning the letter over in her hands, she wondered about why she should feel obligated to serve a system of justice that had given her no openings. She resisted tossing out her summons, in a satisfying gesture of refusal, only because she knew she

FIGURE 4.1. The front steps of Tina's trailer, May 2, 2019. Photo by Nicholas Shapiro.

would face consequences for doing so, even if the state's own rules safeguarding her survival would not be enforced.

I met Tina in the spring of 2019, in the high desert an hour and a half north of Los Angeles. It was almost fourteen years after Katrina, and some 1,800 miles from the Gulf Coast. I'd heard people ask versions of her question for a decade at that point, from Maine and Saskatchewan to Arizona, Florida, Alaska, and most states in between. Bundled up in her succinct *"Who* is responsible for this?" are a host of other questions I've been asked over the years: How can I return this unit and get my money back? How can I receive some sort of restitution for my family's health issues? Who will hold accountable the companies that made these trailers, the government that commissioned them, and the resellers? How can we ensure this doesn't keep happening to other people? The paths to realizing any of these four desires—those of refund, injury compensation, punishment, and deterrence—inevitably routed through the courts or, at the very least, through stiffly worded letters from lawyers. Yet the solutions promised by these tools remained mostly out of reach, as they were for Tina. Even if some kind of restitution materialized, the realities of the legal process melted claimants' expectations for justice. Although it was too much to relate in that moment, and I couldn't yet put it in the words I have now, I had recently begun a halting reckoning with how the illusory state of justice for people in situations like Tina's were bound to larger fantasies of the rule of law and the utility of punishment (fig. 4.2).

This chapter grounds its pursuit of these expansive questions with the stories of those seeking answers and justice after living in a FEMA trailer. It extends well beyond those bare-bones homes to interrogate the profound shortcomings of the figures and processes of everyday accountability in the United States that are generally recognized to be the avenues for remedying past harm and deterring future harm.

To start addressing both Tina's and my own questions, I'll focus not on the uneventful fringes of the resale market, but rather on the spectacular aftermath of Hurricane Katrina, where the possibility of justice was closest at hand. This was the site and the moment where lines of responsibility were clearest, exposures the highest, and media attention was most focused. Where the constellation of homes was most spatially and jurisdictionally concentrated and the timelines of inhabitation were most synchronous. All told, upward of ninety thousand people stood together with dozens of lawyers to demand recognition of and to seek redress for their chemical injuries. The ubiquitous, chronic, low-velocity, and invisible formaldehyde exposures that span the nation's domestic spaces were most visible in these moments. This chapter begins with the stories

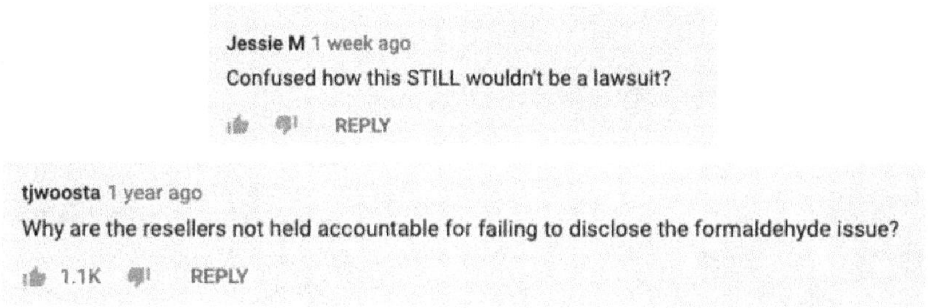

Jessie M 1 week ago
Confused how this STILL wouldn't be a lawsuit?

👍 👎 REPLY

tjwoosta 1 year ago
Why are the resellers not held accountable for failing to disclose the formaldehyde issue?

👍 1.1K 👎 REPLY

FIGURE 4.2. Comments on YouTube regarding a mini-documentary the author made with Mariel Carr about resold FEMA trailers in 2015.

I tell to people like Tina, when they approach me after being locked out of the traditional avenues of justice and accountability—stories of circumstances far more conducive to traction in the courts than their own (fig. 4.3).

The scenes that follow outline not just how the courts failed to advance justice for the chemically injured, but also how failure was the anticipated outcome, to some extent, of most FEMA trailer inhabitants with whom I spoke. Almost no one that I met was brimming with hope that their participation in this product liability case would bring justice, let alone survival. The litigation—which stood as the only plausible avenue for recompense or advancing their welfare—became a small gesture of disappointment long before it ultimately concluded in a paltry settlement.

The superlative eventfulness of the issue is undeniable. The failures of this case to render anything like recognition, restitution, or accountability indicate the limits to Elizabeth Povinelli's incisive and important assertion that "late liberal modes of making die, letting die and making live are organized within and through a specific imaginary of the event and eventfulness" (2011: 134). Her observations help to explain the difficulty of cultivating political intervention into slow, corrosive chemical exposures that don't thicken into conventionally perceptible events. But this relationship between eventfulness and life, enervation, and death perhaps better accounts for the production of abandonment than it does the possibility of intervention and change. In other words, the case at hand helps elucidate how uneventful wounding may explain more about how people are made or let die, than eventful wounding facilitates justice or protection. Even an exposure heard around the world was not sufficient to propel meaningful change. There will never be enough wounding for the late industrial world to reconsider its relationship to toxicants like formaldehyde.[2]

FIGURE 4.3. The author in conversation with a mother whose son and grandson became sick while living in a resold FEMA trailer in Oklahoma. Binders of her research cover the table. Photo by Akasha Rabut.

This is not my theory, assembled from fieldnotes and scholarly reading. Rather, it is theory taught to me over and over by those in the thick of it.

Far from thinking that their quickly infamous exposures would yield either recompense or welfare, far from denoting a coherent attempt at restoring impaired rights, those whom I spoke with largely perceived their participation in the lawsuit as more perfunctory than justice-oriented. To the extent that they expected or encountered resolution, it came in the form of letting go of fantasies of legal accountability and moving on, moving life forward without reference to the exposures of the past. But many of those who were still overwhelmed by the severity of their illnesses five or more years after moving out didn't have the privilege of turning the past loose. Their exposures clutched to their every movement. For these people, who were always on the brink of complete exhaustion, the possibility of imagining survival required that they find ways to physically remove the past, whether through surgeries or other body care.

Beyond the figure of the heroic lawyer and the promissory rhetoric of mass tort lay feelings, actions, and theories of the world, enacted by the plaintiffs themselves, that more closely resembled the triage of survival and recording their demands and experiences for history. I'll start detailing these receding horizons of what torts really do with three stories from people whose participation in the litigation came with very high stakes, yet each knew that the actual results would hold little to no bearing on the future. These three stories evince both a spoken and unspoken recognition of the vacancy of litigation's promise. After outlining the limits of this promise, I then turn to a small opening in its monopoly on popular imaginings of repair. Multiple exposed people articulated the inglorious reality of attempting to heal without achieving their goals of accountability or structural changes, grounded in the reality of enduring the unrepairable (Harney and Moten 2013). This chapter draws deep influence from a postpessimistic approach to "figuring life and responsibilities beyond the individualized body" (Murphy 2017: 497), as exposed people work to find accountability or, failing that, some form of peace.

The Lost Epidemic

I rattled my knuckle across the ajar door of room number 342 with my left hand as I pushed the door open with my right. A focused white doctor was running through some of her patient's possible arrangements for chemo and radiation therapy, following an upcoming CT scan of the patient's latest tumor. After a slight pause to acknowledge my presence, she turned and began discussing a hole in his most recent skin graft.

The patient sat, somewhat delicately, atop a mess of blankets and sheets. His beard hid hollowed-out cheeks, but what you could see of his face suggested that he was probably in his late forties or early fifties. His white legs, and to a lesser extent his arms, looked cadaverous. Muscle, vessel, and bone were visibly distinct to the point that an anatomy class could be taught on his leg as it was, without need for incision. His thin body hardly made an impression on the mattress.

Every so often, the embers of a flinty-eyed ferocity would flicker to life in his otherwise direct, heavy gaze. This mostly happened when he spoke. I could hear the effort it took to push words up and out over his tongue, his enunciation loosening at the end of nearly every sentence. "Basically, I just want what is most likely to work," he told the doctor. He was in the hospital that day to see if his seventh tumor, a lemon-sized protuberance in his left armpit, had metastasized into his lymphatic system. It had. His tumors had started a year after Mac moved out of his FEMA trailer.

As the doctor departed, Daniella, who has been hosting Mac at her home while his house is in the final stages of being repaired, ushered me further into the room. A bob of gray hair matched her trim tweed jacket. Her gestures and words were warm but measured. Daniella and I had been exchanging emails and phone calls for months, repeatedly rescheduling our meeting due to unexpected declines in Mac's health. After our introductions, and without hesitation, Mac launched into telling his story.

When the storm hit, Mac was living in the Mid-City neighborhood of New Orleans, near Bayou St. John, and had already been on dialysis for five years. In the late 1990s his kidneys had swollen up to over six pounds each due to polycystic kidney disease, a relatively common inherited disorder. Both his kidneys were removed, and he was thereafter required to make semiweekly trips to the dialysis clinic to clean his blood. All the while he was still running his wood-flooring business. "Up until the day before Katrina. I was running a business with thirty people," Mac relayed, his nostalgia palpable. "I was running a commando team and knocking out very professional jobs. I've always been in almost Olympic athlete condition. I was in really good shape, and that's why I lived through all this stuff."

After an initial evacuation to Mississippi, Mac bought an old Airstream trailer and parked it on a friend's property in northwestern Arkansas. Mac was in survival mode. "I was on my own . . . up there. . . . All I did up there was keep my trucks going, so I could get my dialysis and make it to my next meal. I just tried to stay alive." After a nine-month stay in Arkansas, he heard word that a FEMA trailer had been set up in his backyard and another had been towed into his front yard for a friend who had no real estate of his own. Mac was eager to get back home and get back to his doctors, as he felt that his dialysis wasn't being handled properly in Arkansas.

"So we get in the FEMA trailer, it's been closed up since we got it, for months in the sun. We opened it up and it smelled terrible. But we figured, ya know, we would just air it out." Mac recounted, "We put fans in there and left all the windows open for a couple days and after that, it didn't smell too bad." When he later tested the air in the trailer, he found formaldehyde levels of 372 parts per billion (ppb), well over three times the maximum indoor air concentration recommended by the US Environmental Protection Agency (EPA).

At the time that Mac moved into his FEMA trailer, he was recuperating from a dialysis-related surgery and attempting to switch to a form of dialysis (peritoneal dialysis, or PD) that he could administer himself from home:

> So here I am trying to recuperate from surgery and I am just starting on PD, and so I do that for how long was it? Maybe six to eight months. . . . I'm not recuperating and things are not feeling well and stuff and then I started getting the weirdest experience: My nose is completely clear— although it would bleed sometimes—my sinus is completely clear, my throat was completely clear. I can deep breathe with no problem, no congestion. Nothing. But if I lift my arm up and down I'd be out of breath and have to rest for half an hour just to recuperate. The only way I can describe it . . . what it felt like to me is that the air would go in but the body couldn't use it or it wasn't getting to the cells or sumptin' was, like, interfering with the uptake of oxygen.

Respiration no longer felt like a nourishing exchange. Its restorative chemistry seemingly inert. The most unremarkable of motions became life-draining affairs.

On one occasion, as he was descending into exhaustion, Mac came home to find Dave, his friend who lived in the trailer in his front yard, sitting in his car in the driveway staring off into space. Mac tried to get his attention. "Dave, what's happening?" he asked through the closed driver-side window. Garnering no response, Mac knocked on the window. He could feel something wasn't right. Without turning his head, Dave loosely held up a single finger to the window as a response, then gradually dropped the finger down onto the automatic window switch. As the window rolled down, he turned and whispered, "I can't breathe."

Dave didn't want to go to the hospital because he had lost his insurance card in the storm and would need it to procure treatment. The few hospitals that had reopened were overcrowded, and Dave did not consider his physical distress to be severe enough to warrant such a hassle. But Mac insisted that he was going to take Dave to the hospital. Mac felt an unspoken urgency in Dave's behavior and ran down the block to where he had parked his car. As Mac pulled up, he saw Dave, who had exited his vehicle, draped over his car like he was being frisked. Half-lurching, half-collapsing, Dave fell to the ground. He died of congestive heart failure, there, in the space between the driveway and the trailer.

A nurse strode into Mac's hospital room in the midst of this narrative, severing the moment. "They are going to do the CTs today, and then I'll send a hard copy to Dr. Armstrong." A few minutes later, after the nurse left, Mac noshed gingerly on some ice—a way dialysis patients quench their thirst without taking in too much liquid volume.

Mac holds the FEMA trailer responsible for Dave's death. Months later Mac suffered his own congestive heart failure while nested among his dialysis equipment in his FEMA trailer.[3] Before either of their hearts stopped, Mac had enrolled in a formaldehyde lawsuit. At that time, the only symptoms he associated with the trailer were his draining chronic fatigue and shortness of breath.

With an indifferent shrug, Mac recounted that he had no interest in seeking monetary redress or, on a more abstract level, checking a negligent federal response or punishing trailer manufacturers. His aims were both further up- and downstream. "As far as I'm concerned, when I first heard about the FEMA lawsuit, I was like 'Aww, I don't want to be bothered with all that,' but then I thought maybe I should just do that even if it's just documenting everything for mankind, or whatever." The utility he recognized in the lawsuit was a function *secondary* to the legal process. Mac saw litigation as a means of staking out his own small quantified spot in the historical and scientific record. His desire was to be rendered a statistic. The lawsuit served as the means to that end, a mode of documenting, aggregating, and making public the dangers of formaldehyde.

Filing suit was not a means of survival or a way to claim the entitlements of citizenship. Rather, Mac set his sights on formaldehyde itself as the elemental building block of his affliction and a potential threat to other or future populations. "I wanted to document everything just so that the public arena had the facts, so to speak. Ya know, the statistical facts of what had happened," he told me. He continued:

> My understanding was that this is the first larger-scale exposure of mankind, anywhere, to formaldehyde. And that they [scientists and doctors] were in denial mode, that formaldehyde had anything to do with it [negative health outcomes in the trailers]. So, I determined that if I don't say anything it will just be . . . if a million people do the same thing as me and don't say anything, then there is a whole epidemic and a whole event that's been lost and so I determined I at least wanted to get on record with all this stuff so that they had the right statistics. Where, in the future they analyze 'em, there would be a whole thing . . . more accurate numbers.

A whole event that's been lost. Mac recognized himself as situated between a unique place in the history of human toxic exposure and a common place within a multitude of similar formaldehyde exposures from FEMA trailers.

His body made up one part of the distributed documentation of this anomalous exposure that, taken collectively, posed the potential to render the human health effects of "low-level" chronic formaldehyde exposure in unprecedented clarity. The lawsuit was his means of coalescing and recording somatic harm for public good. For Mac, litigation was not a means to a specific end (money) but the means to an unspecified future benefit for humankind as a species.

In the spring of 2012, I spoke with a toxicologist that the plaintiffs' legal team had contracted to work on the FEMA formaldehyde litigation. One of my first questions was whether joining the lawsuit could actually impact toxicological knowledge of domestic formaldehyde exposure, as Mac and numerous others hoped. Francesca Daniels, PhD, with a Diplomate of the American Board of Toxicology (DABT) credential, a fast-talking and forceful New Orleanian, clenched her jaw and leaned over her desk as she responded. The diamond-encrusted cross that hung from her necklace swung forward. "Now that's a great question," she began:

> I suggested that we do a real data-gathering survey, which was mine. I prepared health surveys . . . it's an environmental history. That would have made a difference, but they [the plaintiffs' lawyers] did not want to do that. . . . I don't know. . . . They didn't choose to do it. Let's put it that way. . . . So, there is no good database, so someone who is coming in saying I wanted to contribute my [body], that isn't there for this particular group. Because they did not do that [let her collect the data]. . . . So that's a regret, I think. . . . But it [her survey] was never used but if it would have been, I would be able to publish this, but it wasn't . . . so it won't get published. . . . It was just too many power plays amongst the attorneys to get anybody to see—Hey! This is *the* biggest database we will ever have on this type of trailer.

There is no good database. Francesca chalks up the lawsuit's failure to scientifically capture the exposures of the plaintiffs to the myopia of attorneys, who were jockeying to advance their own agendas within the plaintiff steering committee. Such dynamics countermanded Mac's and many others' hope for leaving a statistical legacy by way of litigation.[4]

Enduring the Edge

Mac's understanding of the law's potential role in establishing scientific facts comes from his understanding of environmental law, which itself came from Daniella. Daniella, who sporadically interjected factual corrections into Mac's story from her perch on the other side of his bed, had worked at a university environmental law clinic during the organization's most intense period of litigation against Louisiana's entrenched petrochemical industry. For more than a decade, she traveled all over Louisiana, organizing community groups around environmental issues and supplying them with lawyers they would otherwise not have been able to afford. Ever curious, Mac often tagged along on Daniella's trips to far-flung and chemically saturated communities across the state. "She was like the woman in the movie *Pelican Brief*," Mac recounts with a warm grin, scratching his back by pulling his T-shirt to and fro.

The environmental law clinic that Daniella worked for was hugely successful throughout the 1990s—too successful even. The clinic, which relied on the free labor of law students under the supervision of a professor, was consistently defeating highly paid teams of corporate attorneys and halting pollution-generating production. The Louisiana Department of Economic Development began surveilling the clinic. Louisiana Governor Mike Foster publicly vilified the clinic, calling its lawyers "vigilante" "yahoos" on television. And in the summer of 1998, the Louisiana Supreme Court leveled a ruling that dramatically whittled down the capabilities of all student law clinics to a nearly untenable state (Allen 2003: 102–5). Daniella left the clinic at that point, deeply discouraged about the possibility of using legal avenues to address environmental justice in Louisiana. Regarding the FEMA trailer litigation, Daniella noted in a courtly understatement, "We certainly haven't gotten any indication that there is much hope in the legal arena at this point."

When I asked Mac if he saw the FEMA trailer incident as related to larger chemical exposure issues in the Gulf South, he immediately responded, "Well, of course . . . it's part of the bigger issue because it's part of the chemical contamination of, of . . . the world." Daniella elaborated: "Because of my experience, we see the links and how it's dealt with in the health arena. It's the same situation. . . . The barrier has not really been broken to legitimize these issues in the legal world." *It's the same situation*. Mac has come to very literally embody the chemical political economy that Daniella made a career out of challenging.

At first, Mac felt that the degree of his toxic infusion was minor, but significant, perhaps more so when accounted for with the hundreds of thousands of other FEMA trailer residents. Yet, as cancer surfaced, was painfully quelled, and then resurfaced, again and again and again, Mac's view of the lawsuit-as-archive, a sort of scientific lark taken on behalf of humanity, began to change. Slowing his cadence as his story eased into the present, Mac assessed his position:

> I'm concerned at this point now the formaldehyde trailer has brought me to the point of almost killing me. Put me on the edge. Through this and everything I'm still not feeling right. And I don't know what it's going to do. I don't know the long-term effects of this shit. This is just what happened from immediately after—a few years. What is going to happen in the future from formaldehyde poisoning? That is why I want to find a detox. I want to get this shit out of me. It ain't out. I can feel it. . . . I'm still contaminated.

In four separate moments with me, Mac discussed his lawsuit in the light of his current apprehensions and uncertain future. Each time, his words inflected with distress, he ephemerally appeared to have charged the litigation with greater import as the magnitude of his wounding grew larger with time. It would seem that he was more invested in punishment, compensation, or legal justice. Yet, on every occasion when he appeared on the verge of implying a new salience of the toxic tort proceedings, he stopped short. His outlook from the edge of life had not brought the litigation closer to heart as his personal stakes increased, but rather it fell away from his purview.

The severity and persistence of his injuries had constricted his focus from the future to the immediacy of the harms lingering within his own body. His talk of benefits for our species had receded. Now, Mac had become focused on expelling the cause of illness from his own body rather than contributing to a statistical account of the collective body of the exposed. His hopes were narrowly focused on the alternative medical practice of "detoxing," which he turned to each and every time he contemplated his current situation out loud. "I haven't been able to do it yet, but that is the first thing I want to do. Soon as I can do anything. . . ." The urgency of the feeling that he is "still contaminated," that every biological process deep beneath his skin is laden with the uncanny residues of exposure, attenuated his horizons from the lumbering velocity of progress in exposure science to an anxious monitoring of the ongoing present. He is stuck in a loop of constant self-appraisal. Does he feel well enough to "do anything"? This traversing of temporalities—from long-term scientific change to a preoccupation with the present—breezed past the lawsuit settlements of the near future, with its constricting view. Litigation was beyond his hope, as he and his forward-looking sentiments were utterly exhausted by both the haunting persistence of chronic ailment and eroded expectations for the law.

"I have a lawsuit," Lizzie, whom you met in chapter 2, recalled with indifference. "I haven't talked to my lawyers but once and that was years ago. I don't care." Gazing across the carpeted living room at her third child, who sat askew in his car seat, his neck locked in a perpetual leftward glance, toes at a constant point, she continued, "There is nothing I can do about it . . . except take care of him." Rhett was conceived and largely brought to term in a FEMA trailer in Pass Christian, Mississippi. His gestation would have been fully housed by the trailer if Lizzie had not fallen ill and had to be hospitalized from vomiting any time she entered her home.

Rhett was—to a large extent—a healthy baby. Yet his legs would from time to time turn blue as a blueberry, leaving little purple polka dots when the spells had passed. Sometimes he would also stop breathing. According to the doctors, they could hear the blood flowing normally through his arteries with their stethoscopes—"shwoosh shwoosh shwoosh"—which made the discoloration all the more mysterious. "It's really intriguing. If you ever find out, please let me know," one puzzled physician told Lizzie as their appointment ended with no discernible leads. About Rhett's occasional apneic episode, they were less alarmed, telling her that "babies do that."

One night while Lizzie was away at a night class for nursing school, Rhett, then six months old, stopped breathing again. Her husband, who was home with Rhett and the other two children, called 911 and performed CPR on the wiry standard-issue mattress—not an ideal surface for the procedure—until an ambulance arrived six minutes later. Rhett was declared dead on arrival at the ER. A determined female doctor—whose gender Lizzie speculated as a driver of her determination—revived him after several minutes of fervent resuscitation. He was documented as having been dead for sixteen minutes, but in all likelihood the actual duration was twice that. Lizzie didn't leave the hospital for six months during Rhett's convalescence. Lizzie and Rhett's triumphant return home turned bittersweet, as her husband disappeared when the two of them got home. He was, apparently, too traumatized to see Rhett alive again.

As we sat in her quiet suburban home, Lizzie began to list the slew of diagnoses that stemmed from Rhett's anoxic brain injuries in her slight southern twang: "Quadroparalytic, cerebral palsy, cortical blindness, drop foot—you see his feet, his tendons are shortened—scoliosis, pharyngeal dysphasia, which means he can't swallow. It just goes on and on. . . . You want a rundown of our daily routine? That suction machine [pointing], I have to stick a tube down his

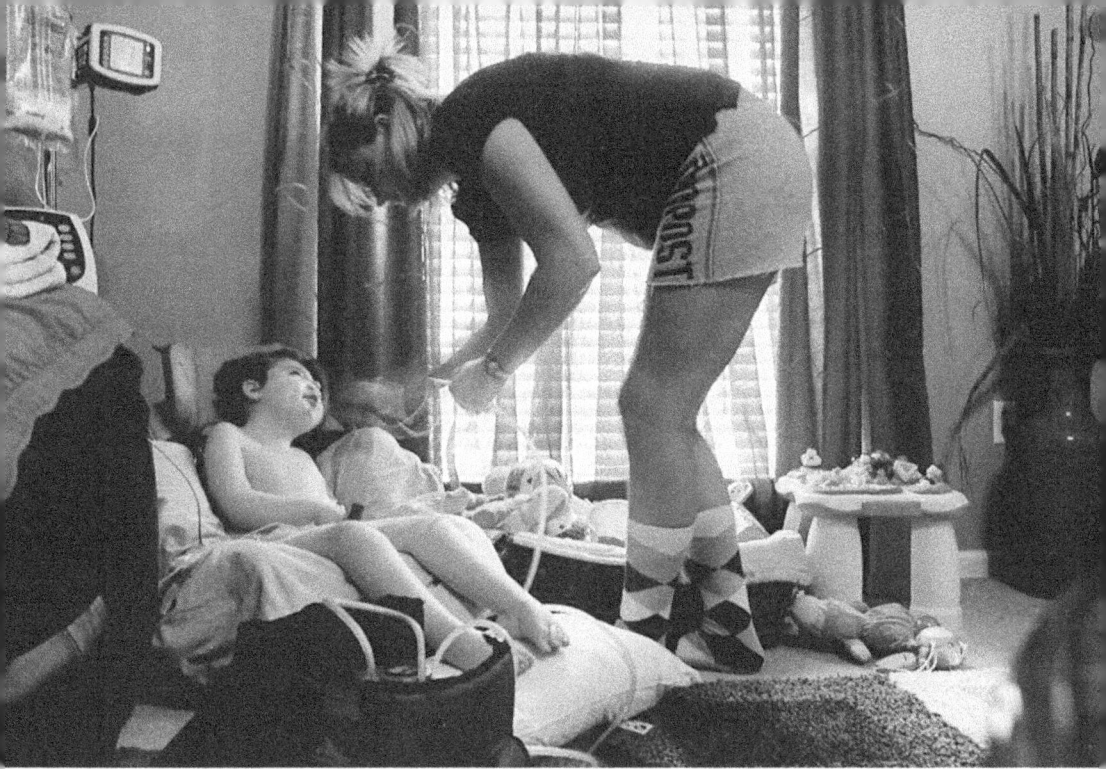

FIGURE 4.4. Lizzie suctioning mucus from Rhett's mouth, June 24, 2011, Southaven, Mississippi. Photo by Nicholas Shapiro.

throat like fifty times a day at least" (fig. 4.4). For a series of months Rhett was seizing seven hundred times a day, then two hundred to four hundred times a day, for two years.

Almost two years after Rhett returned from the hospital, Lizzie rekindled her relationship with the father of her first child. A month after they remarried, Lizzie's husband committed suicide. Rhett's seizures stopped when his stepfather died. "I don't understand the connection, but itnit weird? Not one. They did a weeklong EEG and there was not one," Lizzie explained with a slightly distant look on her face, her eyebrows lingered, slightly raised. "Last time there was 324." The TV Guide channel murmured in the background. The elder of the two girls, the late husband's daughter, was playing on the floor, on the other side of the coffee table. Having overheard our conversation, she responded to the spectral aura by pausing, then reminisced, "Every time he would kiss you his beard would tickle [you]." Lizzie shut that line of conversation down. "Okay, we are not going to talk about that, Christi."

Taking a deep breath, Lizzie returns to the litigation: "I know there's not going to be any amount of money that is going to make up for what happened.

So that's why I'm not so fully involved in all of that. Because I wouldn't care if they gave $100 million out to everyone. . . . So it [litigation] doesn't really interest me much 'cause there's nothing that can really make it okay. There is not going to be an answer. I never will know anything for sure." Rhett's injury leads to an ever-present past. A time horizon that is unflinching to litigation's promise of monetary restoration. Litigation was a gesture of disappointment that was complete upon filing.

There is not going to be an answer. Unlike Mac's original position of statistical futurism, Lizzie sees no imminent knowledge in the future that will make sense of her family's injuries or greater misfortune. Under the weight of biological and economic reality, she does not engage in the forward-looking labor of rendering her story open to future explanation or reasonable redress (Good and DelVecchio Good 1994). She is not struggling to justify hope or pine for fantasy solutions, as she has come to the tacit awareness that monetary or scientific hope is more trouble than its worth. Lizzie's focus is on pushing back against the constant past-becoming-present of Rhett's disabilities, a struggle that now also takes precedent in Mac's life. While both Mac and Lizzie have enrolled in toxic tort litigation, their signatures on registration documents belie their lack of hope in the litigation's outcome.

In June 2012, Rhett's two front teeth fell out. Lizzie posted on Facebook that she broke down crying—it was an all-too-literal kick in the teeth. Just another disfigurement, one among the many diminutive horrors he has endured. Mid-sob, she realized him losing his teeth was actually "a wonderful and natural thing. That he is five, and that is what happens when a child is that age." His corporeal relation to trauma and the ordinary is inverse to his brother's occasional eruptions of bodily disturbance in the form of vomiting spells. Although Rhett's body is dominated by a traumatic formation of time, haunted by his past death, on rare occasions normative measures of time unexpectedly break through, conjuring hope and the counting of blessings. Reactions conditioned by the need for life-saving hypervigilance melt into unanticipated pockets of joy.

Despite the omnipresence of family illness that she attributes to the FEMA trailer, Lizzie has pushed her interest in formaldehyde into the past. She recounted a phone call with a friend that occurred just before I came over to her new home in a suburb south of Memphis in the summer of 2011: "My friend just asked me what I was doing, and I told him I was waiting on you, and he was like, 'For what?' and I was like, 'Formaldehyde stuff . . .' He was like, 'FOR WHAT?!' He don't even know. I know him, and he has no idea. I don't mention it. It consumed me for a really long time, researching and trying to figure out why and what and how and what could have been, and I had to let it go." Rhett started to gurgle. Changing her tone, she sprung up and said, "Here you can actually see." Lizzie jumped around the coffee table and began to suction Rhett's throat. "Cough it up! COUGH IT BACK UP!" She ordered with increasing sternness. The mucus that was causing him to gag whipped through the clear tubing of the suction device. With relief she lulled, "That's good, Rhett. You needed that."

The Toxicity of Tort

Before I met Mac, Lizzie, Daniella, and the others who taught me about the practical and experiential irrelevance of legal accountability, I volunteered at one of the law offices that sought redress for injuries from living in FEMA trailers. The lawsuit targeted three classes of defendants: the federal government (which supplied the trailers), trailer manufacturers (who made and sold the trailers), and trailer installation companies (who installed them in a way that increased water intrusion and thus mold and increased formaldehyde off-gassing).[5] The corporations that were contracted to install these temporary housing units as more permanent residences—such as Shaw Environmental, Bechtel, Fluor, and CH2M Hill—were summarized by one attorney as "Halliburton-esque" in terms of size, close ties to the federal government, hydrocarbon extraction, and the logistics of war.

At the office, my responsibilities were somewhere between that of a law clerk, an intern, and a consultant. I assisted with preparing deposition materials and evidence lists. I researched defendants' expert witnesses and met with plaintiffs' expert witnesses. I worked under a high-minded attorney, Rikki Elver, who in turn worked for a less perceptibly high-minded attorney, Steve Nibbs, who ran the firm. As he rhythmically spun the oversized trackball on his computer mouse while he interviewed me for the position, it appeared that Nibbs was less concerned with my actual interest in the case than in making sure I was not a mole for the formaldehyde industry. Nibbs, a short clamorous man whose office sign is in the shape of a flexing white arm, vastly thicker than his own, is the sort of lawyer who has a rhyming slogan. The type that advertises on billboards along accident-prone stretches of the highway—the passive mode of ambulance chasing ethically admitted by the bar. His private bathroom, which we all used when he wasn't around, contained paintings of partially clothed women draped over gothic stone architecture. As many misgivings as this breed of lawyer may conjure, their cash keeps mass tort cases running. Nibbs had invested, I was told, upward of $2 million in the case (although court documents attest to just under half a million dollars in claimed trial costs).

The cost of litigation adds up: years of hired expert witnesses, depositions, travel, attorneys' salaries, paralegals' salaries, constant FedExes, law clerks' salaries, office space, and an almost insatiable demand for basic office supplies. Unlike individual cases, in which the plaintiff usually bears the burden of funding litigation, plaintiffs typically enroll in mass torts without any up-front costs. As a result, the attorney's final cut is likely higher, and mass torts are potentially high-yielding investments.[6] With the investment his own, Nibbs's pathway to profit involved working the minimum number of hours needed to win a hefty settlement. Nibbs's firm was just one of nineteen law firms on the plaintiff steering committee that represented upward of ninety thousand individuals in total.

Rikki, my direct supervisor, was nearly always exhausted from the long hours he worked on the case. His burly Midwestern frame oscillated between hunching over binders in his office and pacing the hallway while talking on his Black-Berry. When I would catch his ear, he would emerge from the nuts and bolts of litigation with rhetoric rife with grand concepts of justice. Despite Rikki's passionate drive, it became increasingly evident that the firm as a whole was involved with FEMA trailer exposures to make money and not necessarily to carry out the broadly conceived notions of "justice" that Rikki saw himself as purveying—two motivations that are not necessarily mutually exclusive. Just two examples among many: On one of my first days, someone working on another case, on another floor, called out the sound of a cash register opening, "Chaaaa-*ching*," through the office intercom after reaching a lucrative settlement, likely due to someone else's injury. Before quitting, a law clerk informed me that she was asked to go through the plaintiff listings and roll back the dates of initial consultation so that more attorney hours could be logged. By the end of my tenure at the firm it was patently clear that the plaintiffs were positioned less as "clients" to be served than as intermediaries between lawyers and their cut of a multimillion-dollar settlement.[7] This doesn't categorically preclude the hope for justice that both Rikki and I held tight, but it did appear to make it incidental to the desired outcomes.

Beyond the competing incentives of counselors, the impersonal and lumbering process of mass tort litigation was discouraging for plaintiffs. In the eyes of the law, this product liability case wasn't as straightforward as, say, a single product built by the same company—like a faulty pill or an exploding phone—and it failed to be certified as a class-action lawsuit. The trailers were built by dozens of different manufacturers, and, despite the toxicity of formaldehyde being invariable across brands, the court ruled that there was a theoretical diversity in the mechanisms of injury, precluding class-action status. The litigation proceeded instead with a series of bellwether trials. Bellwether trials, an increasingly common phenomenon in US tort law, are cases taken to court as (ostensibly) statistically representative of plaintiff subpopulations.

In the lead-up to one of the early bellwether trials, the defending trailer manufacturer opted to settle, as it was well on the way to bankruptcy. The bellwether plaintiff, who had already spent countless hours spilling the private details of her life in depositions and the discovery process, was not entitled to increased compensation for serving as a bellwether case due to the defendant's bankruptcy. Another law clerk described to me the sheer desolation of this plaintiff. She steadied the woman as she staggered to her car, tears drying on her cheeks. "The check for a couple thousand dollars, if that, is going to be a slap to the face," the clerk said, her gaze ruefully skimming the floor.

In addition to the opportunistic involvement of the firm and the distress brought about by the process of mass tort itself, plaintiffs were increasingly disheartened by the legal complications encountered by their lawyers. The government, which is the entity most responsible for the damage in the eyes of most of those with whom I spoke, was cut loose from liability. First Louisiana, and then Mississippi, and then Alabama courts dismissed the federal government as a viable defendant due to a lack of "subject matter jurisdiction." More precisely, the courts ruled that FEMA could not be held accountable for any damages resulting from providing housing units due to sovereign immunity conferred to the state during an emergency. The Federal Tort Claims Act provides that "the United States shall be liable in the same manner and to the same extent as a private individual under like circumstances" (28 U.S.C. § 2674). FEMA, then, is shielded from analogous private liability through each state's emergency management laws, which protected those supplying free housing to the displaced as Good Samaritans. Legally, the courts ruled, FEMA had no more responsibility to provide housing than well-meaning bystanders with a spare room. Thus, FEMA had no liability. The court's findings underlined the predetermined state of exception that accompanies emergency, flattening normal hierarchies of governmental duty and liability (cf. Fassin and Pandolfi 2010).

When I last saw him, Rikki had been relocated to a diminutive cubicle in the attic of the building the law firm had moved to after a fire had ruined their old offices. The fire itself was suspicious; some employees speculated that Nibbs himself had been involved due to the damage being allegedly centered around his files. The rumors reflected the clerks' and lawyers' smoldering anger at what they considered Nibbs's critical underpreparation for the trial. Rikki's parting words to me, a paraphrasing of the head of the plaintiff steering committee, were that the FEMA trailer suit would remain nothing more than "enduring badness." He left the firm for a job with a firm involved in litigation related to the BP oil drilling disaster. He encouraged me to follow him and recalibrate my work to focus on oil and dispersant exposure. I almost did.

As the enthusiasm and righteous language ebbed, and defense-friendly rulings rolled in, it became increasingly clear to me that justice, redress, accountability, and future deterrence would not be what lay at the end of the proceedings.

Land of the Litigator?

I can remember, with an uncharacteristic clarity for that time in my life, when the notion that Americans are hypervigilant injury litigators burrowed into my brain. I was a quiet fifth grader, nestled in the corner of a basement classroom in Atlanta. The teacher had a collection of confiscated Tamagotchis jangling on his belt. It was unclear if he despised these handheld digital pets or delighted in spending his idle moments squinting at their tiny screens and tapping away at their buttons, keeping the contraband digitally alive. I was bad at putting my stool away, and in an idle moment the teacher quipped that he was going to sue me if he tripped over it and hurt himself. A know-it-all ten-year-old chimed in to inform me that that kind of thing happens all the time in America. "You can get rich if you get hurt a little," he noted. The teacher nodded and affirmed my peer tutor's insights with a story about an elderly woman who spilled scalding-hot McDonald's coffee in her lap and was awarded nearly $3 million in damages.[8] The case had been settled for a few years, but it was new to me. After that passing moment, I started noticing grifty characters in sitcoms embellishing their fender-bender injuries by wearing unnecessary neck braces, hoping to cash in. The coffee case itself was mocked on shows like *Seinfeld* and *Futurama* and in a song by country superstar Toby Keith.

I was happy to now be in on the joke. These cheap laughs landed over and over again because of a growing common sense in the 1990s that corporations were constantly at risk of extortionary frivolous lawsuits. Companies like cigarette manufacturer Philip Morris had funneled over $16 million in 1995 alone to advance tort reforms that would limit the liabilities to companies producing hazardous products or services (Covington and Burling 1995). Industry bankrolled ostensibly grassroots groups under the banner of Citizens Against Lawsuit Abuse, and the hot coffee case in particular became a serialized rallying cry from the "Republican Revolution" led by a congressman, Newt Gingrich, from my home state. It was no coincidence that I was learning that corporations were being victimized by the greedy people they injured in Georgia and that "Oh, are you going to sue me?" became a running joke to downplay minor injuries on the playground.

Yet the truism falls short of veracity. It's a false assumption in multiple directions—most injuries are not litigated, and people sue for lots of reasons. A number of quantitative studies, conducted around the time the myth was being passed along to me, indicate that some 90 percent of those sustaining injuries with tort potential don't bring suit (Danzon 1985: 19–21, 23; Abel 1997: 448–50; Nader 2002: 203; Jain 2006). This trend is especially pronounced when consumers receive a defective or injurious product. Today, and for the last five years, tort cases have steadily held an extremely minor fraction of the state civil court caseload: 4 percent (Center for Justice and Democracy 2020). The data on this are a bit dusty, but an old American Bar Foundation and American Bar Association study established that, of American consumers who found major purchases to be faulty, only 7 percent consulted a lawyer (Curran 1977: 146). As legal scholar Richard Abel summarizes, "The tort system does not encourage fraud or display excessive generosity but fails to . . . discourage unreasonable risks" (1997: 447). In other words, the possibility of litigation does not prevent "unreasonable" risks precisely because such litigation is predictably undersubscribed. The myth of overzealous injury litigation is likely an inversion of reality.

Qualitative researchers have also noted social pressures that discourage the filing of injury claims (cf. Engel 1984). In her ethnography of a New England courthouse, anthropologist Sally Engle Merry (1990: 3, 172) found that her interlocutors, largely white and working class, availed themselves of the law with much reluctance and only after other means of mediation were found ineffective. This is all to say, Lizzie and other like-minded former FEMA trailer residents are not alone in their apathy toward leveraging the law in pursuit of injury restitution. Not only is the American will to sue often a last-ditch effort of a vast minority of injured parties, but none of my interlocutors who had filed suit seemed to hold fantasies of landing a windfall via their involvement in the case. Prior to and during the start of my tenure at the law office, I had bought into the prospect of a lucrative payout and possible punitive damages, but the dozens of FEMA trailer residents I spoke with held no such delusions. It didn't take legal or toxicological knowledge to know that the process of accountability or restoration would not be found in the courtroom.

These industrially cultivated injuries enhance income for lawyers litigating the harms, the healthcare system treating the injuries, and the corporate

manufacturers that cut corners to enhance profit. The injured parties themselves become collateral in an injurious system of value generation (Jain 2006).[9]
Shortly before the paltry settlement offer in the spring of 2012 (each claimant
would be awarded less than $250 on average), a lawyer in a white suburb north
of Jackson, Mississippi, frankly explained the process to me during an exploratory
conversation about me serving as an expert witness for a case they were thinking
of taking on. "Come on, these kinds of tort are not designed to pay plaintiffs. If
and when they win, lawyers get paid, not the plaintiffs," he stated blankly and
released a stream of chewing tobacco into an empty Mr. Pibb soda can.[10]

Even before numbers were on the table, the FEMA trailer inhabitants I spoke
with were aware that the numbers that would eventually come their way would
be dismal, if they materialized at all. They were aware that the case would
likely settle for a pittance and that the defendants would admit no wrongdoing. Most were, in essence, just going through the motions of justice, which
did confer some sense of conscientiousness. Some, like Mac, initially joined in
hopes of extending scientific knowledge about chemical harms. In June 2013,
while hospitalized for breathing difficulty, Mac received notice that his claim
had been evaluated to be worth $228.07. He passed away peacefully later that
summer. Rhett's lungs collapsed for the final time in late 2012.[11]

As previously noted, state courts categorically dismissed claims of governmental malfeasance. Around 75,000 to 90,000 cases (plaintiffs' lawyers acknowledge that even this range is speculative) have been settled for meager
sums, none of which involve criminal charges or acknowledgment of civil mis-
conduct. But what is more important to me is that most of the plaintiffs had
not, by and large, invested their participation with hopes of enfranchisement,
let alone survival. Some, largely mothers, rejected the civil suit's rhetorical
equivalence of money and bodily damage. Like Lizzie, they did not buy into
litigation's promise that money will compensate for wounding, that recompense bore the fantastical capacity to return the injured to a prior state of life.
Many individuals with chronic health issues opted not to sue. Even more with
ailments that appeared to be temporary chose not to seek redress. More still
didn't file suit for fear that FEMA would confiscate their trailer, a widespread
anxiety among those who had little else.

The bulk of the settlement checks related to the FEMA trailer product
liability lawsuit were delivered by the end of 2013. Debra Coleman of the
Gentilly neighborhood received her settlement check of $32.02 in early No-
vember 2013.[12] She received another check for the same amount a week later.
Her second check was made out to Debra McDonald. Her husband, Clarence
Coleman, also received two checks—the second one made out to Clarence Mc-

Daniel. A dozen friends of the Colemans also received second checks bearing an incorrect last name. Among the recipients, speculation abounded that such checks were part of a fake check scam designed to elicit sensitive information from recipients. The Better Business Bureau urged those who received duplicate checks to call FEMA directly to report and correct errors (Truong 2013).

There are two telling, and surreal, slippages in how the public and plaintiffs responded to the resolution of their legal claims. First, they suspected the very substance of redress and justice—the settlement checks—of being agents of fraud. That the process of recompense for biological damage and social suffering is indistinguishable from a systematic swindle is revealing of the status of tort law in the United States and the expectations of those who resort to litigation. The medium of restitution was suspected to be an insidious agent of harm, like a minor echo of the life-saving housing unit that had shuttled toxic chemicals into their bodies.

Second, the Better Business Bureau assumed that FEMA was the party responsible for administering redress for harms resulting from the inhabitation of FEMA trailers. This connection is seemingly straightforward. However, early court proceedings had released the agency from any FEMA trailer liability years before a settlement was reached. The idea that the government *should* be responsible, that the state will parse acts of assistance from acts of harm, is an assumption that clung to these trailers from their inception in the chaotic aftermath of the hurricanes to Tina's unanswered question (which opened this chapter) some fifteen years later.

On multiple occasions, FEMA trailer residents would—after I had discussed my university-based research in detail and they had read consent forms adorned with my university's logo—introduce me to neighbors, friends, or colleagues as a federal government investigator. The institutional context to which they felt I *should* belong, or was perhaps easier to conceptualize and relay, overwrote the actual context of my work. After I explained to an elderly mobile-home park owner in Texas, as we sat in sun-bleached folding chairs, that I did not in fact work for the government, she began telling me stories from the early days of the mobile-home industry. She recollected a specific incident, nearly a half-century earlier, when government agents drove out to her park to replace defective piping in certain mobile homes. This residual optimism of some FEMA trailer residents, mobile-home park owners, and the Better Business Bureau in regard to the responsibility of government is what Lauren Berlant would call a "normativity hangover" (2007: 286). In the haze of this hangover, nostalgic forms of normativity—that retrain optimism about governmental commitments to accountability—coalesce with simmering suspicions of the state.

Healing Without Accountability

We must keep coming back to the question of what justice looks like to those who have been treated unjustly.

FLINT, MICHIGAN, RESIDENTS IN THEIR COMMENTARY ON A
PROPOSED $600 MILLION SETTLEMENT

In the United States, it's difficult to escape the idea that justice looks like a paycheck. The idea that money will serve as redress to the aggrieved, retribution to those responsible, and a financial threat to deter future harm. But profound limits to how the courts imagine justice and accountability are also difficult to avoid in the long hereafter of exposure. It is not only when the dust settles and the payouts are more readily understandable as scam than recompense that impacted communities find themselves asking what justice could and should really look like. Even if there had been more zeros in the settlement, what has been taken from Lizzie and Daniella—let alone Mac and Rhett—cannot be restored by cash.

As Eve Tuck and K. Wayne Yang (2016) have observed, this place of countervailing currents—where justice is necessary but deferred—is a common end point for "justice" processes established by the state. As "mechanisms designed to facilitate justice almost always come up short" (Tuck and Yang 2016: 5), the tens of thousands of plaintiffs in the FEMA formaldehyde case are not alone in the space between grief and grievance (Cheng 2001; Best and Hartman 2005). This cycle of desire and disappointment can feel stuck on endless repeat, even if each volley of demands helps to shift the terrain of redress and exposure to harm ever so much more toward accountability.

Unrequited demands for justice have been disproportionately borne by people of color and theorized by Black, Indigenous, and other scholars of color (Cheng 2001; Kelley 2002; Best and Hartman 2005; Tuck and Yang 2016). And, indeed, while all races of former FEMA trailer inhabitants came to similar conclusions about their lack of expectation for this civil suit, Black plaintiffs seemed to most squarely face what their lawsuit couldn't do.

Betty, a young Black grandmother undergoing chemotherapy for a cancer she developed after her time in a FEMA trailer, told me as we finished talking over lunch, "I was let down [by the government], yes, I felt like that. I did. But . . . I'm just grateful that I am out of it, you see. I'm like this, you can hold anger for so long, and then you've got to turn it loose—simply because it will consume you. You just gotta move on." She looked out the window, adjusting the green baseball cap that covered her hairless head and matched her dark green pants and blazer.

Within a few months of moving into her FEMA trailer, Betty came down with pneumonia for the first time in her fifty-year-long life. She remembers her physical descent, closing her eyes and putting a forkful of food back on her plate. "I was feeling bad, feeling worse and worse. I was unable to talk for any length of time without getting short of breath." Eight months after moving in, Betty found ballooning lumps under her breast. Since then, she has been enduring the constant hurdles of surgeries and chemo regimens in the battle against persistent breast cancer.

She saw government recklessness as a direct threat to her life, but turning loose her anger about their endangerment didn't mean simply letting the parties that may have caused her cancer off the hook. For Betty, part of the process involved delegating a flawed pursuit of accountability to her attorneys— another white personal injury firm—and periodically calling to check in on the case's progress. Turning loose one's anger or resentment required the security of community and did not preclude accountability.

Betty insisted that, for the defendants, "it's a responsibility that they have, toward each plaintiff. Because I do, in fact, say that there is some responsibility, if not all, on their part. But I'm not gonna walk around with it," raising her eyebrows at the idea that holding anger is her social obligation. For her, releasing her anger toward those that put her in harm's way is what happens when you "have some backbone and stand firm and believe that you have to go further. Once you step further and you go into another phase of it, you don't hold on to anger," she said, nodding in the passenger seat of my car, her hands folded in her lap. Her relationship to accountability was functionally the same as Lizzie's and Mac's but also searing in a different way. She was not just letting anger go out of triage, like Lizzie "having to let it go" to care for her son or Mac pulling back from conceptualizing justice because he "could already be dead right now and not know it" and wanting to focus on remediation.

Betty asserted her agency in squashing her negative emotions toward her exposures and her health, as she knows better than to, as she put it, "sit down and wait until the government comes in and handles it." She knew not to wait for incrementalist "progress" of chipping away at the white supremacy enshrined in the law (Baldwin [1963] 1992: 100). Her survival and peace in the face of all of the horrors of displacement and the predictable anti-Blackness of racial capitalism are a resistance beyond the comprehension of the courts.

Simply committing their stories to this page is also a small form of justice, a refusal to forget, like Mac's desire to be recorded, that exceeds the haphazard logs of the litigation. But changing the microconditions of one's relationship to injury and the anger stemming from that injury simply should not have to be the prevailing path to peace. As even torts themselves remain under conservative scrutiny—a perennial desire to pass laws that limit compensation of the aggrieved—reflecting on what justice for injury should look like is more urgent than ever. If reparations are a "construction project," as Olúfẹ́mi Táíwò (2022: 4) asserts, and not simply an adversarial conflict waged in dollars and cents, then the question of "What should we build?" is one way of moving closer to justice. The next two chapters take seriously calls to build the world differently. These construction projects differently theorize, through practicing, the politics of interventions and change. They begin near where Mac left off, with remediation.

FIVE

Working the Stopgap

It is a powerful idea to think of all of us as litigators, putting
the world on trial, but does it actually work? Do the
material and political wins come through?

EVE TUCK, "Suspending Damage"

Drive east out of New Orleans. Over Lake Pontchartrain, past a drained
marsh turned suburb, through groves of strip malls. The highway will gradually rise above the flat, waterlogged horizon on cement pillars sunk into the
bayou. Looking out the window, views of hardwood swamp canopies open out
onto the slow muddy current of the Pearl River. At the apex of the bridge's
gradual curve, you transition from Louisiana to Mississippi. Alluvial lands
then give way to firmer ground, and the elevated roadway drops back down to
the earth. Soon you'll pass signs for NASA's Stennis Space Center.[1] Sprawling
across twenty-one square miles of pine forests, this industrial campus is home
to the United States' largest rocket-testing complex. There, in the late 1980s
and early 1990s, among the rocket-test stands and engineers continuously cal-
culating how to achieve escape velocity from Earth, the Stennis Environmental
Research Laboratory developed a model home for the human colonization of
the cosmos, later dubbed "1 Main Street Mars."

Much like the materials that give form to the average US home today, this speculative Martian shelter was composed primarily of engineered materials. To cultivate life beyond our planet's protective atmosphere, the house was tightly sealed, limiting both air and energy exchange. Upon entering this all-white, half-cylinder herald of the future's housing stock, scientists were greeted not by the refined airs of space-age living but a thick cloud of volatile organic compounds (VOCs) that were continuously and quietly being emitted by the home's construction materials. Scientists and engineers who, even briefly, entered this mock home developed burning eyes, burning throats, headaches, and respiratory distress. The very structure that might serve as humanity's refuge after we corroded our planet to an irreparable extent could not extract itself from the industrial conditions that rendered it necessary. Homesick even before leaving home.

NASA was decades ahead of the building industry in coming to understand that reducing air and energy exchange in the name of reduced fuel consumption would yield human health costs if building materials went unchanged. Some two decades before the construction of 1 Main Street Mars, over a hundred VOCs were detected wafting through the crew compartments of the Skylab 3 mission in 1973.[2] Far from being an oversight of a large bureaucratic organization, the toxicity of 1 Main Street Mars was somewhat by design. Its architect knew it was, and even *wanted* it to be, an extra-hazardous use of conventional synthetic materials.

Before living its retirement as an educational tool near Stennis's visitor center, and before it was destroyed by the very hurricane that spawned the Federal Emergency Management Agency (FEMA) trailers, the structure known as 1 Main Street Mars was originally dubbed "the Biohome." Bioremediation practices, first used to clean up the hazardous waste from the Apollo missions that had seeped into nearby wetlands, were being applied to the toxic atmospheres of total encapsulation. Beginning in 1980, Dr. Bill Wolverton led a research group at Stennis, focused on bringing his outdoor water bioremediation victories to indoor air. In other words, this extraterrestrial home was built to be mundanely contaminated and then biologically scrubbed clean. Focusing on formaldehyde, Wolverton's team developed a series of small chamber tests to understand common low-light houseplants' ability to degrade formaldehyde and, later, other additional toxicants. They interpreted their results to indicate that plants were extremely effective at cleaning chemically tainted air. The Biohome was meant to scale up these findings and serve as a proof of concept for developing regenerative atmospheres for space travel and colonization and, later, the common domestic toxicities back on Earth. The home was almost equally human living quarters and plant living quarters.

Drive north from NASA's Stennis Space Center for about twenty minutes toward Picayune, Mississippi. The structures that pierce the pine-tree canopy quickly transition from rocket-testing towers to the spires of rural churches. Just past town is a nondescript mobile home, like so many others, set back from the road and dusted with mildew. Inside, an elderly couple, the Grangers, whom you met in chapter 1, are quietly aging. Their FEMA mobile home bears a footprint almost identical to the Martian model home. And like the Biohome, the chemicals in their every breath are thought to be corroding their bodies and minds, expediting their and their pets' journey through life and toward death.

Between these two Mississippi homes—exceedingly different in use but also materially almost identical—emerged a foray into advancing change and harm reduction without involving the state (as law and regulatory science do). Moving back and forth between the physical prophecy of space-age domesticity and the form of housing that has come to symbolize an American underclass, I began to work on a small-scale bioremediation intervention. This would be my first attempt to practice and theorize a horizon of change beyond those that most readily pervade a dominant imagination of justice.

As methodologist Eve Tuck has concisely noted, "A theory of change helps to operationalize the ethical stance of the project" (2009: 413). Almost the entirety of this book so far has slipped into what Tuck has identified as the unthinking, de facto theory of change of social science research, namely that cataloging harm would compel accountability systems to respond and adjudicate these grievances. In chapter 4, I documented how the primary accountability mechanisms in the United States often heave evidence of environmental harms *not* into a safety net that provides avenues for repair and shifts systems, but into a confusing fog of fantasy and misconceptions. This dominant evidence-of-harm theory of change is predicated on an assumed relay between well-meaning academic research and often nonexistent or nonfunctioning systems of repair and restitution—a handoff into the mist. As Tuck notes, this way of theorizing change is "borrowed from litigation discourse" (413). Ideas of documenting *enough* harm to crest cultural and political thresholds of eventfulness are part and parcel of this "unreliable and ineffective" (415) approach to catalyzing change.

Tuck additionally argues that this hazard- and harm-centered reasoning advances representational harms by pathologizing the lives and landscapes of Indigenous and urban communities of color, leading to damaged self-images.

In other words, this omnipresent idea of what research (or the law) does for the world injects conceptual harms while it fails to make good on change.

To grapple with the scope of discussing concepts this large, I'm breaking this topic into a two-chapter inquiry. As this book moves from working within dominant theories of change to attempting to forge other paths at the periphery, the form of storytelling will also shift. The questions of these final chapters were set into motion by the people that fill the earlier chapters and by the inability of government science and litigation to foster safety or justice for them. Yet, the following pages will not be "peopled" in the same way, as I shift from wading through flawed systems of change and accountability to trying my hand at pursuing the needs impacted people articulated to me. The technical fields I navigate out of a felt responsibility to the people I spent years listening to, and watched being failed, are not being naively hailed as pure solutions. Rather, across this chapter and chapter 6, I attempt to work demands for survival and knowledge that might fuel contestation together with a no-nonsense appraisal of what, if anything, these interventions are good for. The work of this chapter focuses on attempting to rise to the perennial questions brought to me of "How much formaldehyde is in my air?" and, once their bodily insights are confirmed, "What can I do now that I know I'm being exposed and have nowhere to go?"

At every step, the technical work is also concept work, situated assessments of theorizing of change, and experiments in minutely reimagining how the world works. I try to weave multiple, sometimes irreconcilable, theories of change into deeply collaborative efforts to move beyond the cul-de-sac of the methods in which I was originally trained. Along the way I hope to infill some of the no-man's-land between the ways forward imagined by techno-solutionism, critical views of technocratic fixes, and small steps toward survival as a means of treating one's homesickness for a different future.

Theories and Timelines of Change

Before getting deeper into any one way of theorizing of change, let me lay out a schematic of what might even be considered as theories of change. We'll start where this chapter starts: in the stopgap. A stopgap intervention is like holding your hand over a leaking pipe when there are leaking pipes across a whole nation. The leak is stopped, or less intense for the time being, but the problem isn't fixed locally or beyond. At any moment the leak could be back to the way it was, because the root cause was not addressed. Still, it buys valuable time. Respite from persistent destruction secures survival in the short term and also allows for strategizing, building capacity for further theorizing of change, and simply catching one's breath. A stopgap is the fastest form of intervention, often at the smallest scale. It can manifest as mutual aid, like picking up groceries for your elderly neighbor as harm reduction, a needle exchange for injecting drug use, or in the form of remediation tools like the one explored in this chapter. Implicit in community-organized stopgap interventions is not just the support of life in the face of ongoing exposure, but also a denunciation of the state that has failed its charge to steward social welfare (Spade 2016, 2020).

Documenting harm often follows behind or in parallel with stopgap interventions. Documenting harm can begin as simply writing down exposure/harm phenomena in a journal. This is exactly what people did on a fracking field in Pennsylvania where I worked. A middle-aged white woman recounted how her family would document which way they faced in the shower and where rashes later occurred on their bodies to see if they correlated, and they noted the different shades of metallic tastes that would drift through their well water. Folders of notes, like those that Harriett McFeely amassed, were common in the homes I visited. But documenting harm can take many more forms: cellphone videos, documentaries, books like this, or, as I'll discuss later in the chapter, through scientific instruments that perform a certain kind of authority. Documenting harm can support further theorizing of change in a similar way to how stopgaps can give exposed people enough respite from the onslaught of exposure so that they might begin to theorize the patterns of injury. Even if documenting harm does not compel change in the way it is popularly and professionally imagined to, it can be foundationally important for framing what, in fact, the problem is, the key work needed for any attempt to address it.

If litigation is the prevailing form of accountability, policymaking is a predominant way that change is imagined—especially among white-collar workers. Yet, the line between thinking of litigation as a tool of accountability for past wrongs or a tool preventing future harms is blurry. Some tort lawyers have described themselves to me as being "faster" policymakers, as big settlements can de facto regulate markets. Despite some forms being faster than others, the speed of both lawsuits and policy change is often lumbering and slow compared to stopgaps.

Despite their slowness, the scale of impact of litigation or policy change can vastly outstrip stopgap interventions. Ranging from local to national or multinational, the impacts of these government-mediated changes can be large. Due to these similarities, and because of the detailed encounter in chapter 4, I don't depict litigation on its own in this discussion; instead, I subsume it into policy and regulation. This way of imagining change will be unpacked further in chapter 6 along with the concept of nonreformist reforms that serve as a bridge between policy that *regularizes* harm and changes to the structure of society that makes the need for these harms obsolete, or what I call alter-engineering (fig. 5.1). Together, these final chapters will chart this territory from the small survival-focused stopgaps to reimagining some of the very basic structures of society.

More important than any individual project or approach, for the purposes of this book at least, is what they collectively teach us about the self-justifying logics and timelines of interventions. All too often, people who are interested in change pick one or two theories of change to invest in and cultivate some skepticism for others. This makes sense, as these different routes toward the same or similar goals often have discordant logics. Let's think with the bookends of change: Radical systems change (alter-engineering) on its own can hold those presently suffering hostage to a quasi-plausible future that will unfold long after their immiseration and death. Stopgap-only approaches triage out the future, fail to prevent harms, and operate at the smallest scale of intervention possible. Together these multiple logics and temporalities of change provide relays for safeguarding the present while also scaffolding for radical, paradigm-shifting change. Struggles for environmental and social justice need to coordinate the ethics of interventions across multiple theories of change and timescales in order to make stronger strides toward less toxic worlds. Each mode of intervention is individually insufficient, while also necessary.

The kinds of changes that are needed to address the concerns of Tina (which began chapter 4) and the issues that began this chapter are not the same. I don't want to romanticize the way investing in multiple, sometimes contradictory, methods will unfold. The actions of these relays are not those of conflictless handoffs or cleanly siloed approaches simply being put into conversation; rather, their incompatibilities provide productive frictions, forms of contact that both elucidate differences while putting them in touch. Like grist between the stones of a mill, the contact between multiple logics of change is not just a grating loss of energy but potentially rewarding and nourishing, bringing into focus their interdependence so it can be acted on (Hamraie 2017: 99–102).

This chapter begins with a stopgap form of exposure mitigation—the most rapid and smallest-scale intervention possible—for those stuck in the path of exposure. These forms of harm reduction are often the most attractive option to a disarmingly diverse array of stakeholders: those who feel powerless to make big changes, those who need remediation to have the energy to make big changes, or those who hold power and are not interested in big changes. After being inspired by Gulf Coast histories of science to build devices that wield microbial and plant life to continuously break down toxic atmospheres, I turn the analytical attention of this chapter to the medium-term intervention of a very specific form of harm documentation, one that bridges formalized science with community-driven science: building a low-cost tool for quantifying indoor formaldehyde concentrations.

This second project of this chapter *slightly* bends traditional technoscience theories of change of outside actors coming in to document harm, instead imagining building accessible technologies for communities to self-assess risk. Attempting to enable communities' own enumeration is a tiny, maybe even partial, step outside of the conventional roles of science that Tuck critiques. But even this minor bending of research trajectories exceeds the timeline of the fifteen years of research that went into this book. This work illustrates how wide an expanse even just the medium-term entails, preparing us for thinking through the even more protracted timescales of governmental action and the radical reworking of society that will be explored in detail in chapter 6.

Respiring with the Rhizosphere

If you've ever seen an advertisement at a plant nursery (I've seen them everywhere from big home improvement chain stores to local family businesses) or an infographic on your social media feed that extolls the benefits of buying a houseplant to grow clean air, the genealogy of that assertion most likely traces back to NASA's studies in Mississippi. Indeed, when talking to people who were living in FEMA trailers long after Katrina, the leader of the Sierra Club formaldehyde campaign recommended buying a Boston fern to mitigate the withering domestic exposures as a result of this research.

As I began to research this simple, accessible, and cheap stopgap intervention, poring through beautifully periodized images of researchers sporting mullets in acid-wash jeans, its celebrated efficacy began to unravel. The relevance of small chamber studies to the real-life chambers of homes was swiftly criticized. In the early 1990s, the chief of the Analysis Branch at the US Environmental Protection Agency (EPA) Indoor Air Division and later the director of the Indoor Air Division made careful, but damning, critiques of the applicability of these NASA tests to everyday life (Axelrad 1992; Girman 1992).[3] Exciting results from NASA that common indoor plants could scrub upward of 90 percent of the formaldehyde from a half-square-meter chamber melted into impracticality when it came to light that scaling up findings from small exposure chambers to whole homes would require interior forests (fig. 5.2). Some 680 plants

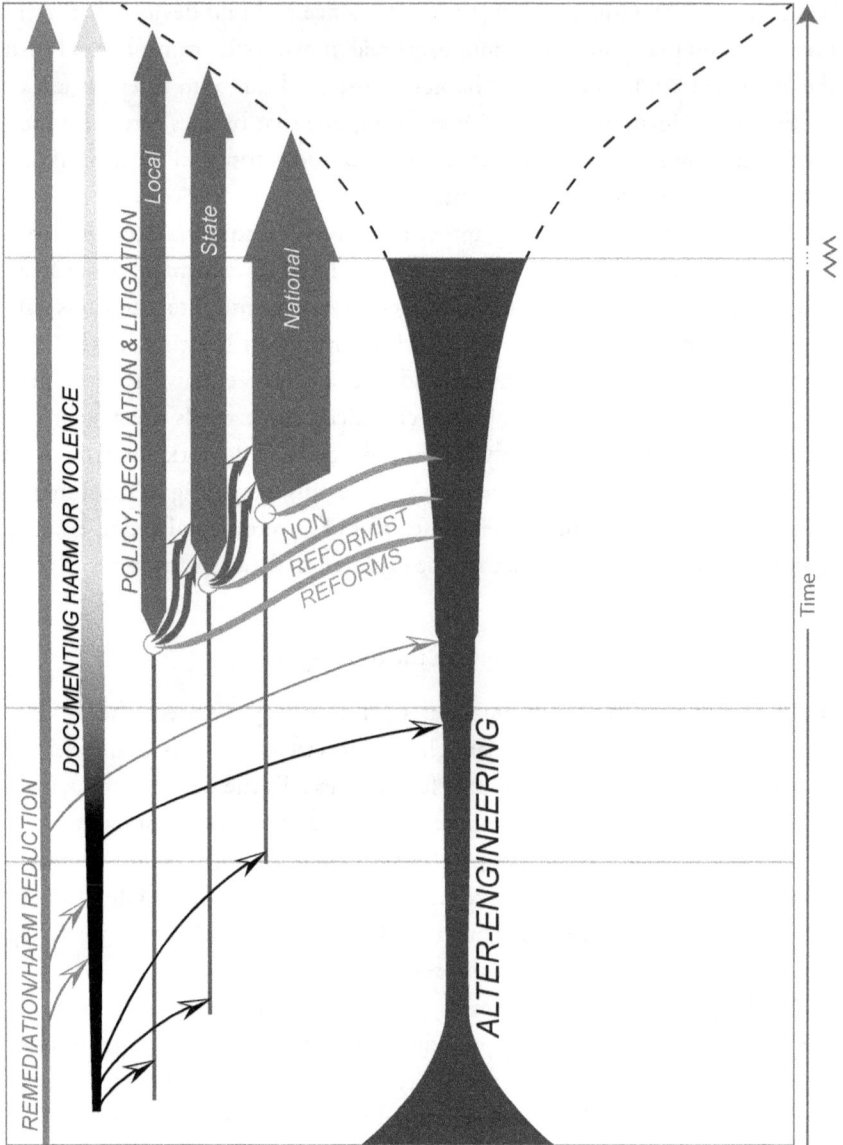

FIGURE 5.1. A schematic of the temporalities and logics of change. The thickness of a line represents the scale of impact. Time moves from bottom to top. The ellipsis on the time axis indicates that the duration of time for alter-engineering projects to reach their potential is unknown. Image by Amisha Gadani and Nicholas Shapiro.

FIGURE 5.2. Images from within the Biohome, 1 Main Street Mars. Images courtesy of NASA. In one image, a container of Colonial brand sugar sits on the table as a commodity link between earthly and extraterrestrial colonial practices.

would be needed to replicate the result within the average US home. Indoor air scientists additionally fretted about the possibility that false ideas of the air-scrubbing potential of plants might lead not only to money poorly spent but to actual negative health impacts from increased dampness from wet soil or illnesses from microbial activity. Even if the air scrubbing were to work and the moisture and microbial issues were addressed, toxic chemicals might be degraded into harmful metabolites, shapeshifting the harm.

Bill Wolverton, the scientist who led the NASA chamber tests, pointed to the Biohome as living proof that his science was not irrelevant to either the colonization of the solar system or the continuing settler colonization of North America. Mass spectrometer and gas chromatograph analyses indicated, he asserted, that almost all of the VOCs that silently emanated from the home's building materials had been removed by the plants he had placed within the structure. Beyond the quantitative assessments, he maintained the decisive importance of bodily knowledge: "The definitive proof lay in the fact that one no longer experienced burning eyes or other classic symptoms of sick building syndrome when entering the Biohome. This was the first 'real world' application of interior plants for alleviating sick building syndrome. A student . . . lived for one summer in the Biohome and experienced no discomfort from indoor air quality" (Wolverton 2010: 4).

Among the palm and fern fronds, which lent a tropical air to the white plastic and metallic interior, was a single experimental potting system that was supposed to accentuate the air-cleaning power of simply plopping some plants into one's home. This system pulls air through the root system of plants because the bacterial communities cultivated in soil by plant roots, known as the rhizosphere, break down chemicals like formaldehyde with much more consistency than the plants themselves. The idea is that these bacteria feast on common human toxicants, and increasing the air flow to the rhizosphere overcomes the inefficacy of the aboveground portions of plants to clean indoor air on their own. In other words, the findings of the chamber studies that scale abysmally were limited not by their metabolic potential but by air diffusion through the plants' less charismatic regions. Pulling more air into contact with the rhizosphere, it was thought, would upregulate the toxicant appetites of these bacteria by, Wolverton (2010) hypothesized, upward of one hundred times.

This low-level cyborg vision of mechanically assisted "natural" processes was deeply alluring. That such a simple system would primarily break down toxics and not simply sequester them like the Brita filter in your fridge was one of the key appeals to me, as low-income families would not need to purchase a filter subscription to survive. Unlike particulate-matter filters that quickly clog up when they are most needed, for example when wildfire ash blows into town, and in normal circumstances need to be replaced at a regular interval, the bio-phyto (bacterial-plant) remediation system is hypothesized to become *more* efficient at greater formaldehyde exposure. Its only maintenance would be that of any other houseplant: sun, water, and the occasional splash of fertilizer from time to time. And also, to me at least, this system felt like it also bore a nondidactic and maybe even ironic lesson in relating and reciprocal cultivation.

To advance this air-scrubbing technology for the benefit of people enduring exposure today, I needed to navigate between the rocky shoals of naive optimism and potential profiteering. The houseplant air-cleaning myth is now a common punching bag for indoor air-quality scientists. Some pejoratively compare the VOC absorption potential of a single plant to the air-cleaning power of a pair of socks. Similarly, multiple companies have built plastic planters that take up Wolverton's focus on increased air circulation. One merely blows air over the aboveground portions of a plant and sports a completely unnecessary space-age-looking hood. This cartoonish application of the concept also misrepresented its efficacy (*Consumer Reports* 2010). Another company works with Wolverton and sells plastic air-purifier pots that more closely resemble his vision. From high-design futurism to unadorned middle America utilitarianism, companies were selling their plastic-pot remediation systems for more than $200 each. This price tag wasn't within reach for me to buy and distribute. Without data to back up the promissory rhetoric of these systems, I was also not willing to suggest that strapped-for-cash trailer residents purchase such products on their own.

I wondered if this straightforward technology could be both further tested and pulled within closer financial reach of those who might need it most. Placing the fan in the pot seemed to be at the heart of the need for customizations that either drove up the cost or, depending on cynicism levels, provided justification for price gouging. So I began sketching out a system that involved attaching a $6 aquarium pump to secondhand pots found at thrift stores (fig. 5.3). I posted my plans on the Public Lab website (Shapiro 2014b). Field-testing this device posed its own ethical and practical quandaries. Should I test this experimental apparatus in the homes of those most impacted, who bear the greatest possibility of benefiting from it, or in, say, an unoccupied trailer, to protect the already exposed from potential unknown further harm? After a long frank conversation with the Grangers, they decided that potential risks from a single plant were worth the potential benefit. I booked a flight down to Mississippi.

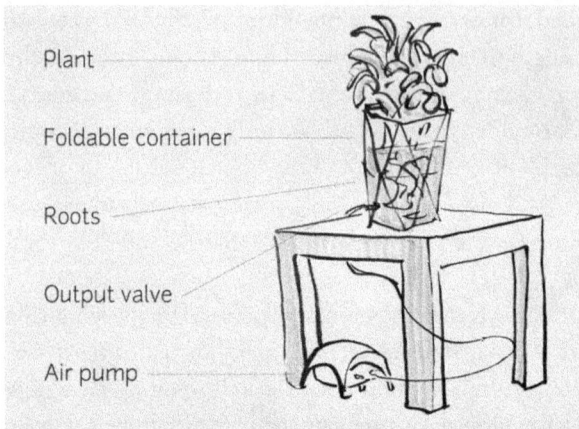

Plant

Foldable container

Roots

Output valve

Air pump

FIGURE 5.3. An early diagram of the plant purifier system with a foldable (and thereby cheaply shippable) pot. Image by Jeffery Yoo Warren (CC-BY-SA).

I worked with a retired NASA engineer named Dan Beavers, who lived just outside Picayune, to test both before and after giving the Grangers my first remediation kit. Between first dropping off the planter and checking in about a month later, we found that the formaldehyde had dropped by half, from 24 parts per billion (ppb) to 12 ppb. Four years earlier, I had found their levels to be much higher: 105.69 ppb to be exact. This information brought some clarity and a host of questions.

Did the remediation system alter the atmospheric course of a toxic home? Did it, in essence, accelerate this detoxification trajectory by about a year and a half over the course of a month?[4] Did the remediation system bend back the chemical-time experienced by the Grangers that had sent their bodies surging toward decay: teeth disintegrating, dogs dying, and skin sheeting off? Or was the result a quirk of context, their air-conditioning bringing in fresh air at a higher rate as the Mississippi spring dialed up in intensity or perhaps more ventilation due to cracked windows or doors? Were fungi and bacteria being broadcast around the home in such a way that might endanger them? Was the apparent contentment of the Grangers with the results enough to keep iterating and attempting to improve the system and repeat the tests with other concerned households?

Even in the framing of these questions it became clear that in introducing an experimental technology, the locus of my attention had at least partially shifted from exposed people and the infrastructures, policies, and sensibilities that maintained their environmental hazards to the cultivation of the technology. The process of seeing if this idea should live or die, of iterating on the parts, processes, and documentation, routed a significant portion of my bandwidth toward attempting to enliven an old idea. The planter emerged as a main character in my daily research life. And quickly I began refining the prototype in my basement that was beginning to morph into a workshop. I moved to the hydroponic growth medium recommended by Wolverton that is thought to increase air diffusion into the rhizosphere and decrease fungi propagation that might differently sicken those in the home. I also added activated carbon to the growth medium to perform some of the standard sequestering (i.e., filtering) of toxicants in addition to bio-phyto remediation.

Another community group that had reached out to me about formaldehyde emanating from the world's largest biomass pellet facility, in South Georgia, was also interested in building a planter based on my documentation. One person placed a plant air purifier that she assembled from readily available parts, and my step-by-step instructions, next to the chair where her eighty-six-year-old mother spent most of the day. Over recent years, the mother had developed large, red patches of irritated skin with well-defined, above-the-skin knots that would often itch, bleed, and ooze (Shapiro et al. 2016). After the planter was used, these signs of illness decreased, in the daughter's estimation, by 70 percent (see photos in Shapiro et al. 2016), and the mother reported that her neurological symptoms seemed to have gotten better in the evenings but worse in the mornings. This lead the woman's daughter to believe she needed to build another planter for the bedroom. Multiple formaldehyde tests from the room where the planter was placed indicated that formaldehyde levels decreased by at least 84 percent after three months.

Over years, collaborations sprawled out to try to make the system more robust and to understand any potential negative impacts. French designer Christophe Guerin approached me after a talk and we worked together to design and manufacture a woven insert that would make converting secondhand pots easier and improve maintenance and aesthetics (it also made them easier and cheaper to ship). We workshopped the design in Paris and Appalachia, while others still experimented with our prototypes in Bengaluru, India; Melbourne, Australia; and Guangzhou, China—localizing the design to work with differently available materials (fig. 5.4).

With environmental engineer Karen Dannemiller and then-students Elizabeth Lara and Allie Donohue, we assessed the microbiome impacts of this air-remediation system with three prototypes of Christophe's much-improved design. Two planters were tested in conventional homes in South Georgia and one in Tina's (from chapter 4) former FEMA trailer in California. By swabbing the dust atop doorways and vacuuming a square meter of carpet into a sterile

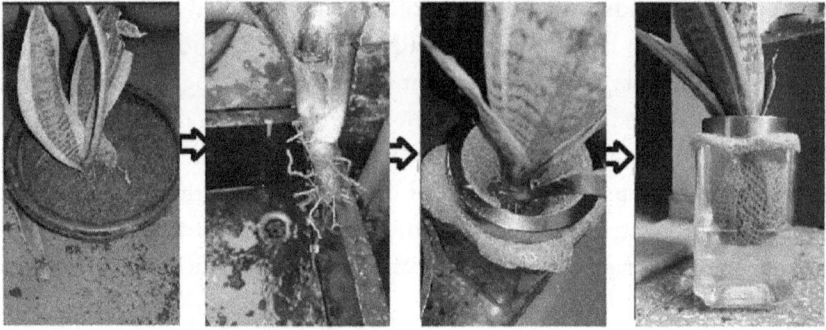

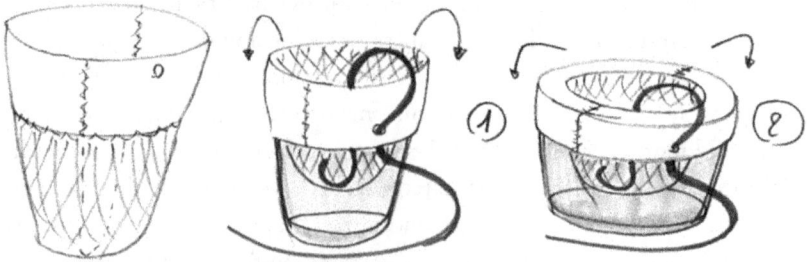

FIGURE 5.4. The top image is a local iteration of the plant air filter in Bengaluru. In the middle are design drawings by Christophe Guerin. At the bottom is an image of building the system with Guerin's design in Guangzhou. In descending order, images are by deeptalaxmibharadwaj, cguerin, warren.

RESULTS

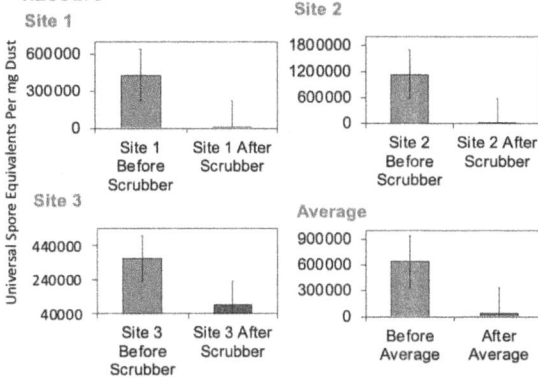

FIGURE 5.5. These charts depict the qPCR fungal quantities in dust before and after the air-scrubber system's implementation for sites 1, 2, and 3 as well as the average of all three. Image from Donohue et al. 2019.

sample filter both before and after planter installation, we were able to determine that across these three sites there was no obvious influence of the remediation system on overall microbial communities in their homes. Surprisingly, fungi levels in the room by the planter were consistently lower after installation of the air-scrubbing system (Donohue et al. 2019), indicating that not only is there a seemingly benign bacterial impact but a potentially health-enhancing fungal impact (fig. 5.5).

This is more or less the state of the project now. Straightforward step-by-step instructions are posted online in multiple languages and openly licensed. It could have been used by thousands or, perhaps more likely, maybe about a dozen. Even finding the site requires a phone or computer, internet access, and, most importantly, knowledge of the search terms that might eventually find the website that hosts this information. The nonprofit that published the documentation has since shuttered and its website now runs slowly and quirkily off a free server in a closet at MIT. This readily assemble-able device of survivability, which can be made for about $20 from items found at big box stores that carpet this county and many others, is running up against the limits of grassroots technoscience (Wylie et al. 2017). Without capital from a large scientific grant or investors that want to start a company, a robust understanding of the efficacy and potential microbiological consequences of the system in different contexts remains out of reach. Grassroots technology development and service provision might also simply be harder for distributed commodity-based communities of use, such as trailers strewn across the county, where there are not preexisting relationships for communication and shared spaces that would facilitate scaling-up impact.

While theoretically accessible because the plans are low-cost, copyright-free, and online, the vast majority of people afflicted by formaldehyde in manufactured homes will not come anywhere close to encountering this information, let alone spend a day running around to get all of the supplies and setting up an ultimately unproven but promising system. Perhaps I steered so far away from potential profiteering by avoiding mass production and commercialization that I failed to make substantial harm reduction. Traditional or traditional-adjacent avenues of capitalism may have mitigated more harm than advancing this tool solely through the avenues of grassroots community science.

The potential of this remediation system remains latent, and it is unclear to me if community members will instigate me to push this forward in the future or if my role was simply to, with incredible collaborators, advance a relay from history just a bit further.

Measuring Beyond the Professional Measurer

Providing free tests to the inhabitants of former FEMA trailers during my doctoral work was already far off the rails of my training in the social sciences. Taking this stride outside of conventional disciplinary bounds, a departure that also enabled me to perform traditional ethnography, had been made possible by anthropologist Kim Fortun in multiple ways. Fortun modeled how to create collaborative research within a discipline based on individual observations and had introduced me to the air analysis lab that processed air samples for me without cost. She even used $500 of her research budget to cover the bulk of shipping costs for the lab to send me test kits. The lab was about to launch its formaldehyde test kit to the public and wanted tests to ensure it was working well in the unpredictability of the field. The idea of somehow building a more accessible testing method was absolutely beyond my imaginative horizon.

The test kit supplied by this private company pulled indoor air into a sorbent material in a glass tube for a short period of time (thirty minutes to an hour) via a small pump the size of a chalkboard eraser. I would then seal the test tube, note down the specifics of the test duration, and mail it back to the lab, where they would run a full-spectrum high-performance liquid chromatography and mass spectrometry analysis on the sample. Each test now retails for $259 plus two-way shipping. As there is the possibility of contamination en route to the lab, technically a trip blank—an unsampled test that is used to indicate if any contamination occurred in the mail—is also needed to ensure the results were accurate. This would push the cost of testing a single location up to over $500. Another option, and the tool that started the Sierra Club's work on the FEMA trailers, is a passive badge the size of a large pendant, which is hung in the house for twenty-four hours before being sent to the lab and analyzed with similar highly sensitive and expensive equipment. This test is around $100 with shipping and, for full assurance, would also need a trip blank. I never encountered a former FEMA trailer resident that had found and used these relatively inexpensive, yet highly accurate, one-time-use tests.

Most people I spoke with had never tested their air other than with their own bodies. A small handful had been given tests by the organizer who had instigated the original community science project (see chapter 3). When former FEMA trailer inhabitants *had* bought their own tests, they were always inexpensive handheld devices that provided continuous "measurement." These detectors, which each can cost less than $20, ostensibly provide multiple measurements across time so that the daily peaks and valleys of exposure can be assessed. The three decimal places of the detector readout lend an air of high sensitivity and definitiveness. Yet, due to cross-sensitivities, inaccuracies at lower levels, and often near-random data outputs (cf. Nold 2017: 105), the fluxes in electrical current depicted on the screen were unlikely to have a meaningful relationship with atmospheric formaldehyde levels. Residents of toxic homes would base large decisions off the whims of these instruments: Should they rip out the walls? Should they cover every engineered wood with a ceramic paint? A nautical sealant? One resident would lay in bed late into the night, trying to derive signals from techno-divination readouts, tools that appeared to be more predatory (false) inclusion into a knowledge economy than indexes of the air.[5] These specious sensors can then lead to purchasing equally murky air-purification devices, like ionic air purifiers that may decrease some airborne toxicants while increasing the secondary formation of others like formaldehyde and ozone via rapidly binding ions (Zeng et al. 2021; fig. 5.6).

Despite prevailing narratives of science and accountability, quantification of exposure alone does not do much for residents. But beyond the wildly erroneous tests, not doing much is still doing something. As my meager research funds ran dry and the steady stream of requests for testing continued to land in my inbox every other day, the small value that I brought to distressed residents also sputtered to a halt. I began looking for tests to recommend that did not spuriously perform precision (like the digital meter) and did not require sending samples off to the lab that required trip blanks or high prices for a single use. I found a method used in the journal *Indoor Air* that had reversed the diaphragm of an aquarium pump to turn it into a vacuum and then pulled air through the inexpensive and already-on-the-market quantification tube (Dannemiller et al. 2013). In the study in which this tool was used, the researchers conducted

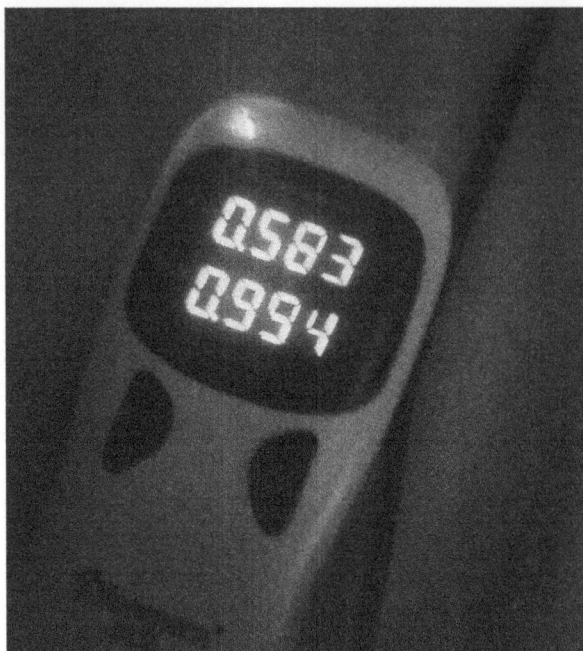

FIGURE 5.6. This reading was taken from within a former FEMA trailer, 2019. The readout of this instrument is mg/m³. Assuming both room temperature (20°C/68°F) and an atmospheric pressure of 760 mmHg, the formaldehyde reading (upper number) is 467 parts per billion, an extremely high level that would likely make a person quickly feel the need to pass out. The lower number is said to indicate total volatile organic compounds (TVOC). Image by anonymous.

seventy formaldehyde tests for approximately $500, yielding an average cost of $7.14 per test. This test was affordable, did not need to be shipped to the lab, and the analog readout performed an approximateness of results more in line with its analytical ability. Employing the aquarium pump instead of the designated highly precise professional air pump saved them an estimated $3,000.[6]

My hope was that I could, with collaborators, publicly document how to make this system, which at the time was behind a publishing paywall, and build infrastructure for sharing the nondisposable parts of the system. I published my idea, documentation, and attempts on the Public Lab website, the same nonprofit that specializes in building openly licensed hardware for environmental investigations where I had posted my remediation designs (Shapiro 2014a). Advancing the momentum of my post, Mathew Lippencott (2015) built a shippable version of the kit inside a crushproof case and, in the process of testing the kit, discovered that the Public Lab kits office bore elevated levels of formaldehyde (26 ppb, 48 ppb during laser cuts). After doing my first tests of the system in my parents' home, I performed more tests in rural Pennsylvania near natural gas compressor stations and in former FEMA trailers in the Bakken shale field. A little less than a year after I had posted my first research note, environmental chemist collaborator Gretchen Gehrke and I were en route to do field accuracy and precision tests of the system alongside the extremely expensive equipment of the Rhode Island Department of Health (DOH).[7] We were testing not manufactured homes but conventional homes that used low-cost laminate flooring that had falsified its self-reported formaldehyde off-gassing rates.

In a data grab and PR move, the flooring company, Lumber Liquidators (now LL), gave out free test badges to customers that requested them. Once mailed in and analyzed, the test-kit result was communicated to the customer directly by Lumber Liquidators. A report I consulted did not state the actual number but simply said, "Based on your test results, it does not appear that your floor is contributing significantly to any abnormal exposure to formaldehyde in your home."[8] The person in Rhode Island who had used their test kit later received a result from a lab with their result of 17 ppb, an ever-so-slightly-elevated level but much lower than all the FEMA trailers I tested (see appendix 1).

When I walked into this reportedly safe home in the fall of 2015, my mind quickly began to curdle. As we prepared the two tests—our low-cost one and the state's expensive and authoritative one—my thoughts were increasingly veering away from the tasks at hand. The pink discoloration, representing formaldehyde concentration, crept farther and faster up the length of the tube than I had ever seen, quickly oversaturating the tube and forcing us to restart the test with a shorter sampling time. We ran eight tests in the house, four with each method. The three tests from our low-cost method yielded an average of 187 ppb, higher than any other home I'd tested. After years in toxic atmospheres, my body's response to exposure had become ever more heightened. As we waited for the last test, I was nearing unconsciousness. My eyelids would linger shut every time I blinked, ephemerally soothing my eye irritation and a rapid-onset exhaustion. Gretchen, the chemist, was not feeling symptoms nearly as acutely. I excused myself, staggering out the front door with shallow gasps for fresh air. I had to lie down in the front yard and passed out briefly leaning up against a tree in such a way that it could look like a normal nap to passersby.

We had been communicating our real-time results as they came in, also noting that the more authoritative results would come in two weeks from the DOH. The resident was clearly, and very understandably, upset. The DOH rushed its analysis of this site's sample and found very elevated levels, but their tests averaged 113.9, a 61 percent error when compared to our DIY tests (Gehrke and Shapiro 2015). The DOH was less than pleased that we had discussed our findings as "very high" even if we had caveated our data as preliminary. This makes sense. "Very high" is not a way that industrial hygienists or atmospheric chemists communicate their results, so my attempt to communicate the acute health hazards, while gritting my teeth and attempting to not freak them out by detailing how it was affecting my own body, was out of step with the department's more conservative and bureaucratic mandate.[9] The tests with the DIY kit showed both high precision and a tight correlation with the DOH tests ($R^2 = 0.99$), but a disconcerting margin of error. High acetone (>900 ppb) and acetaldehyde (>50 ppb) had skewed our results upward.

These moments of validating and calibrating bring into relief multiple issues and tensions. Up front is the worry that I may have misinformed those with whom I had used this test method. The lower the reading, the lower the percentage error so it's unlikely the low readings that the system had been handing out in the oil and gas test sites were providing false distress or assurance. The larger tension that I was wrestling with was that of building experimental knowledge systems to address community needs while also needing to know the capacities and limitations of that knowledge system.[10] I was crushed thinking about the multiple layers of experimental subjecthood, planetary petrocultural experiments, national regulatory experiments, and hyperlocal grassroots scientific experiments. This is a clear example of where good intentions and a critical approach to change and technology are not enough to squarely do good. In trying to respond to a structural problem of lack of access to information, we introduced an individual problem of misinformation. This is a mistake I intend to never make again. Desperate people at the end of their rope deserve better.

Although the system had been published in a top peer-reviewed journal (Dannemiller et al. 2013), the off-the-shelf analytical core was the source of significant cross-contamination limitations. We communicated our results with the lead author of the method that we had iterated upon. Together, with her as the principal investigator, we were awarded a National Science Foundation grant to develop our own color-changing formaldehyde test kit with a price point of $5–10 that would be based on a chemistry that was less susceptible to cross-contamination.[11] We worked with a company that would manufacture and sell the low-cost test kits, and our free and open-source smartphone application would quantify the color change and detect cross-contamination. After a year and a half of application development, exposure chamber tests, and field tests, the engineer at the private company we were working with retired; after the company's factory had been impacted by a hurricane, it ran new market analysis and decided that it was no longer interested in bringing the kit to market. We are, at the time of this writing, working though two Housing and Urban Development grants to advance the technology by hitching it to a domestic allergen and mold detection system, and we are exploring creating a company to shepherd the low-cost technologies through capitalist distribution mechanisms.[12]

While I do not regret my time dedicated to seeing through the development of this low-cost assessment tool, I am deeply unsure if this has been the best use of my time in pursuit of better elaborating the ethical stance for my research. Would attempting to do more conventional science in the unconventional place of trailer parks have been a better use of theories of change that revolve around enumerating harms? While I can't know the answer, I doubt my ethical contributions would be substantially more robust, although the producing of peer-reviewed papers would be more legible to the promotion processes of the academy.

At the same time, I think it's important to also put into context the decision I made as a postdoc that led me to at least a decade of open-source hardware work. Microcomputers like Arduino and Raspberry Pi made DIY computer hardware more accessible in the 2010s, as did steep price drops in computing components. We were experiencing the revival of 1990s dreams of a hackable world, with a specific use case of environmental monitoring (Gertz and Di Justo 2012). Action research and participatory mapping communities were pulled into this promise of community-driven hardware development for environmental justice applications. Epochal disasters like the Fukushima nuclear meltdown in 2011 spurred networked low-cost, open-source Geiger counter development via a group called Safecast (Safecast n.d.). Public Lab, where I was an Open Air Fellow for years, had formed in the Gulf Coast, using balloons, old juice jugs, hacked cameras, and DIY software to aerially map the impact of the 2010 BP oil-drilling disaster on coastal ecologies. Some of these groups attempted to transform the hackathon mode of tech crisis response into a more sustained method of leveraging small amounts of labor by many people to do immense amounts of transparent and openly licensed work.

Outperforming satellite imagery, and collecting data in defiance of corporate and governmental information blackouts, Public Lab had both a regional track record and well-read and acted-upon science and technology studies critiques of black-boxed technoscience to attempt to make technology with better politics. Beyond the accuracy and precision of openly licensed environmental sensors like the one described above, DIY sensors and the other tools of Public Lab enacted something like a MacGyver imaginary. By this I mean a way of reading malleability and agency into a landscape strewn with corporately produced commodities, fanning embers of possibility beyond dictated uses. Many of these DIY tools of quantification end up being more educational and inspirational for broadening interest in hard-to-discern environmental issues than accessible fulcrums for leveraging justice. Such practices also support a historical lens on technology and the environment to see how choices were made, how choices can potentially be unmade or remade, and who can envision and perform this work.[13]

A tool doesn't need to be hyperaccurate and precise to be useful. Unlike my example above, communities need to be accurately informed about the uncertainties of the tools they might use in order to advance the utility of less expensive and less robust tools. Vitally, the efficacy of documenting exposure is determined by not only the tools themselves but also how they are deployed: how many tests were done, what other environmental variables were collected, whether organizers prepared the intended audience to receive the data, and so on. With the right information, study design, and relationship building, noisy data can be wielded toward big victories, such as when charismatically narrated but flawed particulate air-quality data were justification enough for a fracking-impacted community in Pennsylvania I worked with during my postdoc. It then enabled federal scientists to conduct an authoritative study that largely validated their concerns (ATSDR 2016). In this case, years and years of organizing and networking by the impacted community had primed government officials to provide an assessment, and the flawed data were just enough evidence to justify their investigation.

Environmental justice practitioner and sociologist Lourdes Vera refers to environmental monitoring tools that advance community goals without being backed by laboratory-sanctioned accuracy and precision as bearing "civic validity" (2022: 232). Just like laboratory validation, civic validation requires a substantial amount of work to advance community organizing, relationship building, and careful study design to give meaning and power to the numbers yielded by the sensors—forms of labor that may be more easily accessed by impacted communities. But without knowing the epistemic limits of conventional or civic validity of a device or system, community members can experience an acute informational whiplash when devices don't perform as expected.[14]

As I write, Public Lab is almost fifteen years old. The website lives on like a ghost town even as the nonprofit that stewarded it has shuttered. It was a foundational collaborator in this and my previous project. Over the last years, the group had been weathering drops in activity levels and funding. An episode of rebellious, mutual-aid-inspired techno-optimism may have been outmaneuvered by citizen science proliferating across the world via state and traditional science actors in ways that reinforce traditional hierarchies (i.e., using community members for free labor only). Or, antithetically, other community researchers have cultivated a more wary and critical relationship with science and technology and have returned to tried-and-true practices of community organizing. Further still, I might be overexplaining the organization's closure, as it may be minor currents of these factors in combination with the way the COVID-19 pandemic shifted funding patterns and structures.

As the architect Cedric Price joked in 1966, "Technology is the answer but what was the question?" Technology often enacts the question to which it responds. This is especially the case with technologies of enumeration, technologies that count harms in the standard-issue theory of change. With these tools it is assumed that providing new data will provide new political answers, the resolve to seek new political answers, or make new questions askable. If making exposure differently measurable (i.e., less expensive, not requiring lab analysis, an open regime of intellectual property) is the most readily available question to which technology is the answer, what should happen next once someone quantitatively knows they are being poisoned? The remediation system is just one short-term and stopgap response to that question.

The Work Continues

Despite the more than a decade of work pursuing these projects, by the time this book is published the final words of these stories are far from written. In the meantime, I've tried to weigh what, so far, these humble experiments that attempt to materialize (and theorize) slight deviations from orthodoxy tell us about what these theories of change can and can't do in specific contexts. In trying to think forward to the best-case scenario of our accurate and low-cost sensor coming to market, I think of Gwen Ottinger's (2013) work documenting the evolving relationships between the residents of New Sarpy in southern Louisiana and their industrial next-door neighbors. Residents sought a radical solution to their predicament: They wanted the company that operated the facility that was polluting their community to relocate the entire population of their small town. Yet the community lacked sufficient public or political power to force their neighbor to the negotiating table. They used low-cost, community-operated air-monitoring devices so that they could create their own arsenal of data to be mobilized in dealings with their industrial neighbors.

Ultimately the town's goal of relocation was not achieved, but they did use their data to begin negotiations with the company. Their industrial neighbor did offer to clean up some of its emission problems, to pursue more aggressive monitoring (with the help of the community), and to provide financial assistance to community members looking to improve their own homes. While an outside nonprofit urged them to keep pressing for their bigger goals, it appeared that the residents of New Sarpy had seemingly always known that the production of data gave the residents a voice, but simultaneously at the cost of reducing their plight to discrete measurable quantities that could never represent the questions of home, life, and family that were always at stake. Part of this devil's bargain is the way that enumerated exposures constrict the wider, less scientized, forms of homesickness that this book seeks to render.

This use of civic science made the multiply corrosive state of things more bearable rather than substantially questioning it (Fortun and Fortun 2005). And that is also the more explicit logic of the remediation system. In many ways the quantifying step, traditionally seen as obligatory, could be erased by a remediation system if exposed families were to go straight from somatic reasoning to scrubbing their air without quantifying. Instead of relying on scientific "danger zones" that are deeply influenced by industry and often exclude the lived experience of those made more sensitive through exposure, households could simply see if their symptoms dissipate. If they do, they know the likely exposures in the negative, which is a deeply auspicious way of knowing harm. They would not need their knowledge to be passed through the chemical-by-chemical way that industrial chemistry knows the world via its products. I wonder about the effects of skipping the ostensibly required quantification and chemical speciation (identifying which specific toxicant is at play) and going straight to remediation that can break apart a whole host of VOCs simultaneously. These humble experiments in attempting to dilate the possibilities of survival and to expand and reinforce the ethical posture of collective work in turn led to other projects with different timescales and logics of change.

But before moving on to these other ways of imagining and practicing change and survival, sit with the way that this chapter lands with you. Sit with the narrative shape of the stock dramas of science and technology: Will it work? How much formaldehyde were they *really* exposed to? Sit with how the voices of those afflicted become hushed in the hustle of making things, selling ideas, and getting grants. Sit with how hard it is to hold those technical questions—questions aimed at meeting community needs articulated to me—and also hold on to critical understandings of these interventions and also make meaningful change. Sit with the expanses between righteous critique and the material constraints that yield to when trying to put these critiques to work, giving rise to a diluted idea of what the problems are and where they come from. In an echo of Cedric Price, how we imagine (often technological) solutions limits how we conceptualize accountability and the scale and complexity of the problems that drive the harm we seek to interrupt. Chapter 6 digs more deeply into "bigger" theories of change that don't rely on the same promises of technological innovation.

SIX

From Policymaking to
Alter-Engineered Worlds

As I wrote this chapter, I opened the folder in my email where I receive a daily automated alert for all new web pages that prominently contain the word "formaldehyde." That day, in North Philadelphia, formaldehyde spilled in a high school chemistry lab. All the students and staff were evacuated, milling about on the bleachers as a hazmat team cleaned up in full protective gear. A study funded by the American Chemistry Council (ACC) re-analyzed data from a 2010 article. The original article had found that leukemia in Chinese factory workers was correlated with formaldehyde exposure. The new study, which included a 650-word conflict of interest and funding statement, did not find such a correlation. The original study was highly cited in the US Environmental Protection Agency (EPA) Integrated Risk Information System (IRIS) assessment of formaldehyde, and the ACC was using this new study as a fulcrum, attempting to leverage favorable carcinogen risk in the long-anticipated updated assessment.[1] Also, boxer Mike Tyson joked to the press about caringly keeping Evander Holyfield's ear in a jar of formaldehyde after biting it off during a fight decades earlier.

On this day, at a conference in Cambridge, Massachusetts, a postdoc pressed thumbtacks through a poster documenting research on ambient formaldehyde emissions from satellites. The authors of the poster found outdoor formaldehyde levels to be a vital indicator of regional changes to engineered landscapes:

Decreasing levels indicated the efficacy of industrial emission controls in the Houston-Galveston area; increasing levels above the Athabasca and Cold Lake oil fields evinced the ratcheting up of tar sands oil/gas/bitumen production; levels over Illinois and Missouri tracked with corn silage production and likely indicated ethanol fuel production elsewhere in the Corn Belt (du Gouw et al. 2015); and increases in the Pacific Northwest were likely due to reforestation efforts because evergreens, and many other forms of vegetation, emit isoprene to increase their thermotolerance in heat.[2] Once aloft in the atmosphere, isoprene can oxidize into formaldehyde.

A vast array of engineered systems is marked by this single chemical, formaldehyde, on any particular day. From alternative fuels (ethanol), fossil fuel extraction and refining, and high school chemistry to the feed for concentrated animal rearing (corn silage), heat-stressed trees in a changing climate (isoprene), and industrial regulation, the chemical can index much of late industrial life.[3] Any *today* is an arbitrary inroad into thinking through this country and this planet one chemical at a time. Yet, when weeks, years, and decades of todays look much the same, plucking a random day from the history of the chemical is much the same as looking at any day. Squint at this day and it could be more than thirty-five years ago when Anne Gorsuch, the mother of a future Supreme Court justice, was at the helm of the EPA, which was being sued by the Natural Resources Defense Council for failing to prioritize a formaldehyde health assessment that wouldn't be drafted until 2022 and remains unfinalized in early 2025. Industry funded their own studies, leveraged doubt.[4] Some protective regulations were enacted, both then, now, and in between, functioning more as insulations from dissent than health-based policies. Far from garnering the attention of the ABC news helicopter and emergency responders in Tyvek suits and respirators like the slippery-handed high school spill that began this chapter, everyday construction materials privately, silently, and invisibly pour an estimated 54,196 barrels of formaldehyde per year into indoor US air spaces (California Air Resources Board 2007).

The chemical's eagerness to crosslink proteins and DNA, which could have preserved Evander Holyfield's chewed ear so impeccably, also yields the genotoxicity that government epidemiologists project will take 42,000 excess lives. These are the numbers *after* the Federal Emergency Management Agency (FEMA) trailer catastrophe propelled the diffuse toxicity of formaldehyde to the level of international controversy and *after* that apex of attention led to the implementation of landmark formaldehyde legislation in 2010. The Formaldehyde Standards for Composite Wood Products Act, signed into law by

President Obama, was initiated by a Sierra Club petition in 2008 that opened with a description of the suffering of FEMA trailer inhabitants.[5]

This policy is a monumental accomplishment. Achieving national-level legislation is just a pipe dream for most attempts at spurring environmental change. That the FEMA trailers were foundational to policy affecting the breathing spaces of hundreds of millions of people could be seen as realizing Mac's dream of using his wounding to help future populations. It could be seen as the perfect example of why documenting harm, that pervasive theory of change, does in fact work and that Congress can address the issues where the courts fail. For these reasons, dissecting the anatomy of why these rosy assessments are not warranted is valuable for understanding the limits of reformist policy—often the holy grail of researchers and communities alike—and why this legislation fails those with the highest stakes.

In a departure from how outdoor air is regulated, this national formaldehyde rule regulates not how much of the chemical can waft through indoor atmospheres, but how much formaldehyde can be emitted from specific engineered woods that are used to enclose the built environment. As the same composite wood products can be used to create spaces that vastly differ in size and ventilation, and these two variables track along lines of income, it's clear that this commodity-based legislation perpetuates a disproportionate exposure for the poor. Small low-income homes that have little interior room to dilute out-gassing materials and scant ventilation to bring in fresh air are disproportionately burdened by formaldehyde under this legislation. These qualities are superlatively exemplified by the 250-square-foot FEMA trailers, the very spaces that drove the regulation into existence.

As is often the case, this regulation reduces the endangerment of domestic air space just low enough to protect against further public outcry, preserving a sizable portion of exposure that is now likely rendered immune from further regulation. Its harm is optimized. Formaldehyde's regulation ensures its continued place in industrial production, our homes, and our future.

This logic of technocratic harm refinement is not unique to formaldehyde, or even toxics more generally. A lesser-evil theory of reform dominates the popular and technical imagination of what desirable environmental futures look like. From the current attempt to mine our way out of climate change through electrification that requires massive increases in minerals extracted from the earth (Singh 2021) and investments in ostensibly "green" hydrogen fuel that sources its H from fracked natural gas (CH_4) to governmental "biogas" subsidies for large concentrated animal-feeding operations that capture livestock farts and/or burps, the logics of the lesser evil ensure continued planetary homesickness.[6]

In this final chapter, I pick back up the discussion of theories of change, to trudge past remediation and quantification to policy change and radical change. If remediation is a way of asking the question that follows enumeration, radical change requires asking the question before the question of the enumeration harm. It requires swimming upstream from toxicity past the smoking tailpipe, leaking containment pond, or off-gassing synthetic material from which toxicants emanate. Past the durable infrastructures that maintain their circulation. Past the mines and wells from which their precursors are extracted, which might be targets of research scrutiny or policy tinkering. Past political and economic regimes to a somewhat nebulous point of intervention that might be tentatively triangulated by the ethos, subjectivities, and desires that suffuse these engineered systems (cf. Boyer and Morton 2021). I refer to work that intervenes far upstream at the level of not just contesting or diminishing nightmares but building dreamworlds. I refer to work that creates forms of cultural change and service provision that resist being read as "solutions" yet yield a cascade of alternative ways of building, relating, and moving through the world as "alter-engineering." In line with other moves away from antipolitics to alter-politics (Hage 2015) and those that toggle between being with and against technoscience (Murphy 2017), alter-engineering projects do not mobilize counterknowledge to identify tragedies and farces in order to tussle with problems. Instead, these projects seduce our evacuation from the terms that *establish* problems. Not just actively dismantling our engineered world, bolt by bolt, but cultivating alternative worlds that entice its abandonment. Not only locking horns in the unequal adversarial arena of litigation or policymaking, but collectively deliberating the values around which to engineer alternative ways of living.

This chapter briefly outlines the history of indoor air formaldehyde regulation, to make clear the limits of the permission-to-pollute regulatory state. This is not to say that decades of work on regulating formaldehyde have been in vain and that the countless lives saved and prevented suffering yielded by regulatory practices do not matter.[7] Rather, a failure to understand how reformism is just as much, if not more, escapist than radical change has held environmentalism hostage to the diminutive demands that are imaginable through prevailing incrementalist policymaking. My conclusion is that existing regimes of regulation were valuable experiments that have reduced the industrial toxicity of the air, but remaining shortcomings are features and not bugs. The failures intrinsic to detoxifying our breathing spaces from formaldehyde alone are holding back environmental change across the multiple folds of homesickness. In sum, formaldehyde alone is not the problem, and regulating it and other petrochemical toxicants as such ensures perpetual toxicity, lost in a very specific technocratic sauce.

The chapter then pivots to our final logic and temporality of change: radical systems change. If the problems posed by any toxic chemical require interventions and regulations well beyond the molecular encounter of compound and living body, how does the researcher, the afflicted person, or the policymaker move from the already-immense specificity of exposure to an appropriate beyond? The latter half of this chapter takes up this question and attempts to articulate it through the practice of "alter-engineering." This final section may feel disorienting as it does journey outside the empirical scaffolding of much of this book, but also in a way that has been primed since the start. The destination is not singular, but it will be grounded in one such alter-engineering project that I have been lucky enough to be a part of. What I've gleaned from acting on my homesickness for an otherwise future, and reading many others, will take the form more of a series of reflective questions than a specific recipe.

Between Media and Molecule

Understanding the long today of chemical regulation requires, at the very least, tracing bureaucratic engagements with formaldehyde back a half-century. In the late 1960s and early 1970s, a mounting environmental movement and a series of chemical controversies began to expose the cracks in existing environmental policy and stir public and governmental desire for the further regulation of potentially toxic substances. Just five years after the EPA was established in late 1970, the deputy administrator was already noting that their approach to regulatory policy failed to curtail the steady flow of toxicants into human and extra-human environments (EPA 1975). In other words, controlling pollution downstream in collective airs and waters was difficult, as was parsing responsibility and required upstream regulation. The Toxic Substances Control Act (TSCA) was established the following year, in an attempt to grapple with the agency of risky substances as they outmaneuvered legislation that looks separately at air, water, or soil and that draws a line between human and environmental impacts.

Interestingly, the legislation doesn't regulate toxic substances themselves but the commodities that *contain* specific substances. In other words, a substance is not toxic until it is part of something to be bought or sold. As the former director of the EPA's Office of Toxic Substances, Don Clay, succinctly noted, a toxicant "has got to be in something before it can be regulated [by the TSCA]" (Clay 2012). Focusing on commodities is key to the imagined agility of the law, and it also defines its limit. The law's commodity limit was first made evident by way of formaldehyde. As the 1970s transitioned to the 1980s, concerns about indoor air quality in white-collar offices and mobile homes began to coalesce, and new studies increasingly demonstrated formaldehyde's toxicity (Murphy 2006; Ore 2011). The EPA's Office of Toxic Substances nominated the chemical for an expedited regulatory review by labeling it a top priority for the agency. Regulatory scientists recounted the terror and exhilaration of implementing this new law to rein in formaldehyde in 1979: "It was exciting and/or scary depending on a certain point of view" (Aidala 2010).

As the Carter administration segued into the Reagan administration in 1981, formaldehyde transitioned from pressing immediacy to an impossibility. Clay, who would transition from working at the EPA directly to working for leading formaldehyde producer Koch Industries some seventeen years later, noted that regulating formaldehyde with the TSCA "was going to be too hard to do" (Clay 2012).[8] Clay cites two basic issues in his decision to deny indoor air and formaldehyde priority status. The first is that it was too difficult to make the findings—too hard to collect the data across the enormous variation in the types of homes, which were largely made on-site and not built in centralized locations. Second, the commodity limit of the law loomed large as there is no particular industry that produces indoor air. Which industry exactly would be regulated: the architect, the builder, the governmental agency reviewing the permits?

Bearing these factors in mind, and likely also bearing in mind sustained closed-door conversations with the formaldehyde industry (Ashford and Caldart 1996), Clay doubled back on formaldehyde and recommended to the EPA administrator in 1981 that it not be prioritized for regulation. The administrator agreed. Congressional investigations, a lawsuit from the Natural Resources Defense Council, and the dismissal of EPA administrator Anne M. Gorsuch soon followed.[9] To wash their hands of the political attention surrounding the compound and the lawsuit demanding its priority review, the EPA partially reversed course again. They designated formaldehyde as a priority compound, but only for two populations: mobile-home residents and workers in the apparel industry. In so doing the EPA was able to delegate responsibility regarding mobile-home residents to the Department of Housing and Urban Development (HUD) and for the garment workers to the Occupational Safety and Health Administration (OSHA).[10]

The EPA was able to avoid regulating formaldehyde for more than thirty-five years, until the Formaldehyde Standards for Composite Wood Products Act of 2010 went into full effect in 2017. As I noted earlier, the law is inherently regressive, and the small and poorly ventilated mobile homes, which were noted as urgent in the late 1970s, a priority in the 1980s, and an impetus for broad regulation in the 2000s, are the least protected by the law.[11] Governmental projections estimate a reduction of less than 50 percent of cancer cases due to formaldehyde. While the reduction of exposure is significant, so is the remaining risk, which is now further ingrained in the everyday wounding of the country's homes.

The nation's environmental regulatory apparatus inched forward with the 2022 release of the EPA's IRIS *draft* assessment. This cancer risk assessment was initiated in 1997 (GAO 2008). It may one day be finalized and lead to reducing the "acceptable" level of formaldehyde in building materials, which will continue to be regressive, protecting the rich while further enshrining exposure of the low-income, Black, Latinx, and Indigenous people in the law of the land.

Similar to how large chemical companies use global climate change accords to edge out smaller hydrofluorocarbon manufacturers (Tabuchi and Hakim 2016), regulations are an opportunity for market jockeying between chemical companies, and they create opportunities for capitalization. A marketing specialist at the global chemical company Solenis LLC, manufacturer of the leading soy-based engineered-wood adhesive, viewed me as a potential ally in pursuit of tightened regulations to expand their market. In line with larger processes of corporate ghost management of academics to advance their regulatory goals (Sismondo 2018), the company emailed me to make me aware of research they supported that pointed out the limitations of the current commodity-focused formaldehyde regulation. Existing regulations rely on chamber tests of individual engineered woods at 25°C (about 77°F) and 50 percent relative humidity. In their article, they indicate that "atypical" environmental conditions of high temperature and higher relative humidity increase formaldehyde emissions from engineered woods two to three times over "typical" conditions (Frihart et al. 2020).[12] But what are the assumptions defining the terms *typical* and *atypical*? The "atypical" conditions described in the article are often seasonally typical for low-income households that cannot afford air-conditioning or have poor ventilation, such as mobile homes, further indicating how the regulation is a classed and raced license to harm. The company was both right in its assertion of regressiveness and also a fair-weather ally invoking concepts of justice in pursuit of expanding market share and mono-crop soy production in greenwashed industrial production.

Opportunities for profit are not just seized during the process of regulation setting but now also through their enforcement. In 2015, "activist investor" Whitney Tilson of Kase Capital shorted 44,676 shares of Lumber Liquidators (LL) stock.[13] The company's galloping gross profit margin sent the company's stock price rocketing up sevenfold in just two years. These large profits made the company a darling to investors but their advantage in such a tight market raised suspicions to Tilson, who paid for his own testing of LL products. The results indicated that the laminate flooring averaged six to seven times the California state regulations (that were the basis of the then in-process federal regulations). Tilson successfully pitched the story to CBS news magazine *60 Minutes*. The show aired hidden-camera footage of LL suppliers frankly stating that fraudulently tested flooring cost 10–15 percent less than genuinely compliant flooring. The news sent LL stock tumbling by 40 percent in a single day. Tilson reaped $1.4 million in profit.[14] In light of waning enforcement, speculative finance is our regulator now.

Toxicity in this case is not an indication of the sputtering finale of capitalist endangerment (J. Moore 2015: 308) but an opening for opportunistic alliance with the exposed and capitalization. This is likely the other side of the TSCA commodity limit coin. Regulating at the commodity level fails low-income air space but allows for direct conversion of mass harm into financial gain.

Regulatory delay, regulatory capture, lack of enforcement, and a multiply regressive approach characterize the momentous win of getting this compound federally regulated (Dillon et al. 2018). These qualities of chemical-by-chemical regulation are not limited to the United States or just formaldehyde (Boudia and Jas 2014; Hepler-Smith 2019). For example, European toxic laws are often romanticized by American environmentalists, and yet permissible occupational exposure limits for formaldehyde recently increased by 50 percent in the European Union (Boullier and Henry 2022). A sparse informational landscape, partially filled by an industry-funded study written specifically to target these exposure levels, as well as a bureaucratic drive for expediency led to this stark increase in allowable exposures. The idea of the inevitable progress of science and the regulatory state, even in more precautionary Europe, becomes a fallen dream.

As briefly mentioned in the introduction, environmentally persistent chemicals are more easily countable and trackable outside of the lab than those with shorter half-lives, like formaldehyde. In outdoor air, its half-life averages at about an hour (Kaden et al. 2010; cf. Shapiro 2019b). So, it could be argued that the difficulty of regulating the relatively ephemeral compound is partially social and partially the knock-on effects of the unstable nature of the chemical itself. Yet, even a gloss of the history of a group of chemicals known as persistent organic pollutants (POPs), which bear long half-lives and are among the most globally regulated environmental toxicants, similarly undercuts the idea that prevailing chemical regulation practices are progressing toward healthier environments. These chemicals have been subject to regulations limiting their quantities and the most dramatic tools in the current regulatory toolbox: banning and substitution.

As one example among many, polychlorinated biphenyls (PCBs) are a group of POPs that have been globally banned due to their toxicity. One application for PCBs was as flame-retardant additives imbued into furniture. After being banned in 1979, PCB flame retardants were substituted with a set of chemicals called polybrominated diphenyl ethers (PBDEs) that were later found to impair neurodevelopment in children and have themselves been substituted with multiple different compounds that also were found to bear negative health effects (Cordner 2016; Shadaan and Murphy 2020).

This same cascading process of changes that guarantees similar or worse harm has been apologetically dubbed "regrettable substitution." It has occurred over and over from nonstick pans and pesticides to cosmetics, e-cigarette flavors, and food container liners. Environmental scientists themselves are coming to recognize the "intractable, potentially never-ending chemicals management issue that challenges the conventional chemical assessment and management paradigm adopted by society since the 1970s" (Z. Wang et al. 2017).[15]

TSCA legislation was intended to curtail pollution upstream at its basic building blocks. As historian Evan Hepler-Smith (2019) has written, this view of what counts as "upstream" emerges from a system of chemical classification designed by the chemical industry to proliferate chemicals and to align with their inventory and product catalogs. These industry-supported classification systems that privilege chemical structure over how they react and metamorphose became the conceptual framework for regulations like the TSCA.[16] In other words, the net being used to rein in chemical destruction and protect health was built by industry for a nearly opposite purpose: to organize commodity production and order markets for getting chemicals out into the world. With this inappropriate foundation, it is no wonder that the nation's most advanced chemical regulations resemble the intentionally infuriating children's game of whack-a-mole. In this game, nearly identical "enemies" infinitely pop out of nearly identical burrows, unfazed by their reprimands that only ephemerally appear effective, yet still each blow remains somehow affectively fulfilling.

The environmental policy landscape is strewn with changes that guarantee sameness: regulations that enshrine harm to the very people whose suffering catapulted the regulation into existence; regulations only enforced by opportunistic speculative finance; and the most extreme action of banning that leads to differently toxic substitution. What is touted by some as the missteps of progress can also be seen as the deep-seated escapism enmeshed in the so-called realism of dominant approaches to toxics regulation. This calm, suit-adorned, yet also white-knuckled, attachment to processes and concepts with well-rehearsed shortcomings begs for alternative approaches to thinking about where pollution starts and how it should be intervened upon.

Alter-Engineered Worlds

"You can't argue with free," a former industry chemist implored me to bear in mind after a talk I'd given about indoor formaldehyde. He was referencing how methane—a precursor to formaldehyde—when captured in the course of oil extraction, is essentially free and that such a low monetary cost might outweigh any human or extra-human environmental cost. Far from the demoralizing political economic fact that I assume it was intended to be, such a statement helps to reframe how the problem of formaldehyde cannot be addressed through chemistry alone. The truth is that this one chemical, or any other of the myriad petrochemicals, is materially tied to petrochemical supply chains of a world structured by hydrocarbon extraction. Any desire for a future without toxic homes requires a scale of change vastly exceeding the usual informational battlefield of a single toxicant or family of toxicants.

The connections between human health and planetary geophysical stability accidentally outlined by that chemist in no way lead to the magical thinking that simply reducing CO_2 emissions will automatically reduce the corrosive presence of formaldehyde. A material substance is not the crux of the problem of either accumulating domestic formaldehyde or global greenhouse gases. As Matthew Huber has argued, "Oil is better understood as a social relation"; it is a product of both "the materiality of petroleum and the social projects that channel its biophysical capacities in particular ways" (2013: 4). Decarbonizing the grid without grappling with the assumptions, imaginaries, extractive logics, asymmetrical financial benefits, regimes of consumption, and traditions of ecological mastery that were honed in tandem with hydrocarbon engineering allows these social practices and infrastructures to churn on with new "green" lifeblood. If only the material and not its social relations are terminated, the problems of oil, and therefore formaldehyde, are not on a path to resolution.

Both issues of climate change and human toxics exposure are ensconced, not just in material and infrastructural connections but in similar responses elicited by reformers (namely, efforts to minorly mitigate and thereby optimize existing emissions). Over a decade ago, a friend of the critical theorist Eyal Weizman joked to him that a large sculpture of the number 665 should be erected as a monument to the dominant calculative political credo of the lesser evil. By the lesser evil, he meant the prevailing technocratic assertion that some core violences in society are necessary and therefore unavoidable, and moderating these necessary evils is the only reasonable course of action (Weizman 2011). Just one imaginary unit of harm away from the undeniable wrongness of 666. Indeed, from carbon austerity to toxics reform, this is where our dominant environmentalism lives; 665, or the maximum allowable harm, represents the optimal operating emissions level of industrial infrastructure and the basis of permission-to-pollute regulations. Although what counts as just shy of the number of the beast changes over time and space, activists and titans of industry often come down on the same side in terms of supporting the framing logic of necessary harm mitigation. Small wins and well-meaning regulations that emerge within sanctioned adversarial arenas of contestation further root both toxic emissions and their social projects in the future. Similar to the previous chapter, so much energy is poured into the question of how much, of the line between 665 and 666, because it can be difficult to even conceptualize what other questions could and should be asked.

Making harms seen as intrinsic to the current function of global economies unnecessary is almost unimaginable within this commanding logic. The powerful common sense of the lesser evil makes thinking beyond its perpetual triage unreasonable. Within these narrow imaginative horizons, its regimes of urgency make robust alternatives easily dismissible as unrealistic (Whyte 2012: 67). Known violences consistently prevail over proposed cross-cutting pragmatic interventions.

The "alter-" in the title of this chapter does not turn its back on the vital movements that arise in denunciation to oppressive infrastructures, institutions, and relations but additionally argues that enacting alternatives is a vital means of denunciation through affirmation of an otherwise (see Hage 2015). In other words, alter-engineering is saying no by saying yes to practices that the violent norm makes unimaginable. Alter-engineering is not simply de-engineering. It is not dismantling bolt by bolt the most egregious modes of pollution until we have optimized our own global despoilment. It seeks interventions into what presuppositions compose the frame of the problem to be solved. If the problem is a set of norms and not how much of chemical x is emitted by process y, then the intervention is not limited to the calculative logistics of reformism that often lead to flourishing remaining forever just out of reach. Pursuing ever-thickening documentation of chemical harms remains important in the short term, but limiting the conceptualization of the problem and the subsequent action to specific chemicals in specific places often lands communities and researcher allies on a path to justice that takes the form of a data treadmill (Gould et al. 2015; Shapiro et al. 2017).

From Nonreformist Reforms to Alter-Engineering

More important than any whiz-bang technology, captivating policy, or charismatic slogan is collectively establishing values that guide possible ways forward. Value-driven processes for decision-making is what "solutions" fundamentally look like. Traditions of Black and Indigenous feminist organizing have long made clear the importance of understanding process as the ultimate product (Herrmann and Stewarts 2001; Indigenous Environmental Network 2018; CLEAR 2021), but it remains difficult to divest from the deeply downstream location that technocratic capitalism instructs is the proper place to hang one's hopes for betterment. By way of example, Critical Resistance (CR), an abolitionist group founded by Black feminists, asks a very simple set of questions to assess their support of potential actions such as, "Will we have to undo this later?" (Critical Resistance 2021). The question is devastating to the investments of dominant environmentalism. The answer to the "bridge fuel" of natural gas in the early 2000s, of electric automobility of the 2020s, of battery-intensive electrification would all be "yes," if questioned by this rubric.[17] They would all be nonstarters.

This question is part of CR's larger operationalization of André Gorz's (1967) concept of nonreformist reforms, which underline the value of actions that enhance survivability now and capacitate radical change that may be farther down the line.[18] CR works on prison abolition and, through decades of struggle and study, adapted these guiding questions for addressing how to engage in policymaking in the now while working toward full systems change. CR and other intellectual allies worked to make clear that prisons as discrete entities are not the root of the problem, but rather a manifestation of a set of capitalist interests empowered by forces of discrimination (Critical Resistance, n.d.). This approach gave me concise language for understanding that the material harms of formaldehyde that can be monitored and quantified cannot be remedied by thinking about formaldehyde alone, or even by thinking of the bigger tent of petrochemicals.

Another long tradition of standing apart from the common dilemma of critique without paradigmatically different alternatives are Indigenous visions for environmental justice. Indigenous researchers articulate exacting genealogies of violence and recenter ongoing projects of (settler) colonialism as equal and coconstituting forces alongside climate change and industrial pollution. They *also* carry and advance robust traditions of environmental stewardship, accountability, and protocols for being in good more-than-human relations (LaDuke 1999; Liboiron et al. 2017; Simmons 2017; Simpson 2017; Estes 2019; Gilio-Whitaker 2022). Their proposals are not speculative but empirical on a timeline and scale that can be difficult for settlers to fathom. Despite the homesickness of settler colonialism, land held or managed by Indigenous peoples has on average stewarded biodiversity better than land held or managed by others (IPBES 2019). These lands host some four times their share of the world's biodiversity compared to the scale of territory currently under Indigenous control (Sobrevila 2008).

The Red Nation, a coalition of Native and non-Native activists, educators, students, and community organizers advocating for Native liberation, released *The Red Deal* in 2021. The book explicitly endorses the tactic of nonreformist reforms and also outlines bold goals, focusing on "treaty rights, land restoration, sovereignty, self-determination, decolonization, and liberation" (Red Nation 2021: 12).[19] As I finish this chapter, Chile just failed to pass a new constitution that enshrines many of these demands into the backbone of the nation, expressly establishing Chile as a plurinational state and codifying Indigenous rights to autonomy and self-governance (Holton 2022). The fact that such a radical proposition made it to a national stage and accrued 38 percent of the vote is indicative of how close radical change is to cresting the surface of the actual.[20] This is the kind of massive, collectively deliberated intervention that swims up the tributaries of violence in the conceptual feeder streams of homesickness.

The multiple headwaters of a homesick present are a key point of confusion among environmentalists. This confusion can make value-setting more difficult for environmentalists than, say, prison- industrial-complex abolitionists who have forged a much more consistent picture of what must be refused and are establishing the infrastructures necessary for germinal collective dreaming of alternatives to take place.[21] Another way of saying this is that disagreement on the "true" causes of massive environmental destruction has led to differing ideas of what should be done. This is potentially most true among academics that have variously named alliances in the ongoing -ocene debates (capitalocene, plantationocene, anthropocene, chthulucene, coloniocene, and so on).[22]

Each offers crucial and partial insights. Such is the power and limitation of any standpoint, yet there is also disproportionately high value in the standpoints that have been systematically excluded (Collins 1990; Harding 2004; TallBear 2014).

Akin to the multiple logics and temporalities of change, holding this multitude of variously critical genealogies together, with none taking precedence over the others, is vital and tricky work. One place to start is knowing that there is no singular headwater to the modes of relation that put into motion massively scaled harms. Another is tracing out the overlaps in what would need to change to prevent the patterned harms across these genealogies. This would be a small step toward shifting some critical thought from forensics (which all too easily slips toward 665) to coming to understand what these violence tracings share in terms of how to live differently. This may also help to capacitate the both harder and easier question of how we can render ways of making the world less toxic that are not fully predicated on a genealogy of harm (empire, oil, industrialism, capitalism, scientific imperceptibility, insufficient regulation, specific toxicological pathways, uncharismatic agents of violence, and so on).[23]

Some Inroads to Alter-Engineering

There are many inroads to addressing the above expansive question, and here is one that helped me.[24] A shared thermodynamic imaginary (Moe 2015; Engelmann 2020; Engelmann and McCormack 2021), or a conception of how energy does and should work, unites both petrocapitalist worlds and dominant visions of renewable energy. This common vision of energy supports the accumulation of greenhouse gases in the atmosphere *and* the plan to effectively mine the planet out of climate change emissions: minimizing carbon emissions by maximizing mineral-intensive electrification and battery manufacturing. Both problem and reform share an energy imaginary.

In this vision of a sustainable future, solar energy, as an example, is made to work like oil. By this I mean that nearly omnipresent, ephemeral, and feeble solar energy is cast in the form of oil: relatively scarce, dense, extraction intensive, and easily transportable.[25] If only the material (hydrocarbons or their downstream product, formaldehyde), and not its social relations and conceptual roots, is terminated, the problems of oil are not solved. In this light, the chemical industry scientist who told me I couldn't argue with free as an argument to why things would stay as they are is accidentally articulating a justification to radically reengineer the world around meeting energy where it is (Konsmo and Recollet 2019), which is ultimately much closer to free. To continue with solar energy, its power can be leveraged to meet earthly desires without treating it like oil and without the mineral demands and energy loss of conversion to electricity and transportation. It is no accident of history that a world built around a practice of energy that is scarce also features common notions that social power, too, is scarce rather than abundant (Tuck 2019, cited in Liboiron 2019b).

Let's stay in the stratosphere but also ground this very theoretical inroads in specifics. Commercial airlines are a useful starting point to think past mainstream sustainable futures as they will likely never be electrified. Their energy demand and the weight required by contemporary batteries simply defy aerodynamics. Intervening in this space where dominant visions of sustainability run out of steam, Argentine artist Tomás Saraceno has been leading a project to realize aerosolar balloon travel as a viable means of global mobility. I've had the luck of collaborating on this project over the last few years.[26] Aerosolar aircraft are simply large envelopes that are heated by the sun and move with the wind. As opposed to the aerodynamic force that holds aloft planes and helicopters, the aerostatic lift of these balloons does not require concentrated forms of energy to create relative motion over an airfoil (a wing or propeller blade). Rather, the soft architecture of these aerostats becomes buoyant by the expansion of the solar-heated air within it.

Although aerosolar balloons do not require batteries, they are not free of the need for energy to keep them aloft after the sun sets. On multiday flights they remain aloft at night because they are designed to use the infrared radiation that the earth returns to the cosmos at night (think of a piping-hot sidewalk at midnight in the summer). Instead of disposable energy reserves small enough to be portable (batteries), the project thinks on a cosmic scale with our planet itself functioning as an omnipresent source of reserve energy after the sun has set. The shelf life of this means of nocturnal buoyancy is approximately five billion years, or until our sun begins to die. In the thermodynamics of this project, the battery finds its obsolescence deep in the past, before life on Earth began.

Conceiving of wind and solar rays as critical infrastructures, resembling some Indigenous views of clean water as critical infrastructure (Huson and Spice 2019), for the ongoing present demands that our desires be reengineered through largely forgotten supply chains: the planet's shared and circulating atmosphere. These currents of interest are not merely the aerial flows that propelled a nautical yesteryear but multiple, overlapping, dynamic, and sometimes countercurrent atmospheric strata that have only recently become modelable in their full complexity. If you were to look up from this page and levitate vertically from your chair up through the air column, you would encounter many different wind directions and velocities as you traveled through the roughly ten miles of navigable atmosphere up into the upper stratosphere. With different directional and speed opportunities at different altitudes, adjusting one's altitude becomes a way of steering an otherwise haphazard trajectory.

Collaborators at MIT's Earth, Atmospheric, and Planetary Sciences department have developed a web platform that helps to imagine how this unrealized technology could take them where they want to go. This tool calculates when to fly and at what altitudes, in order to string together the right winds and reach a destination of their choosing.[27] Such an engagement with both the elements and big data evinces a project with a complex orientation: both of and beyond this world.[28] Just as with any large-scale transition, making fossil fuel–free and battery-free travel imaginable is foundational for cultural, fiscal, and technical shifts.

Figure 6.1 depicts two long-distance stratospheric balloons on display at the Grand Palais in Paris as part of the COP21 Solutions conference, where an ancestral version of this chapter first circulated. The image takes us back to the iron and glass of the arcades that performs a technological mastery over wind and rain while embracing the sun. These very building technologies epitomize the visual splendor of an early version of industrial capitalism that sought to overcome the environment.[29] If a member of Saraceno's studio had not been stationed in the rafters, inflating and deflating the aerosolar sculptures based on the fluctuations of sunlight, the aerostat would have Willy Wonka'ed through the 115-year-old roof and buoyed up beyond the "green" cars on display that are somewhat visible in its metallic reflections. This ultimate symbolic rupture could very well have happened when the person responsible for monitoring pressure and temperature in the balloon was nearly arrested by Parisian police.

In any case, the way this project theorizes change is not through the empirical dream of evincing harm into oblivion but the realization of the transit infrastructure latent in the complexities of our everyday ecology. This approach to change also manifests a foundational ethos for a mode of alter-engineering that does not attempt to overcome or interrupt nature in an eternal quest for mastery and comfort but attempts instead to realize a human relationship with the environment predicated on calculated submission—using cutting-edge tools and data to move with the elements rather than overpower them (Engelmann and McCormack 2018; Boyer and Morton 2021; also see Calvillo 2018). If racial capitalism produces "social separateness—the disjoining or deactivating of relations between human beings (and humans and nature)" (Melamed 2015: 78, drawing on Gilmore 2002), then projects that aim to overcome these produced divisions are working on making racial capitalism obsolete.

FIGURE 6.1. *Aeroscene* installation at the Grand Palais, Paris, for the COP21 Solutions conference. Photo by Nicholas Shapiro.

The aerosolar project draws on a reflective nostalgia (Dlamini 2009) that pulls centuries-old technologies, cutting-edge science, and the fabulations of science fiction into the same frame. As is often the case, disparities in revenue sources undergird where the most follow-through can be directed. Some members and supporters of the project can get frustrated when developing infrastructure, technology, and community take a backseat to communicating the potentiality of the idea in art exhibitions. Over the last five years, the project has built longer-term collaborations with Indigenous communities on salt flats in South America to try to correct a mismatch between gesturing to a future and building it. These communities are facing the prospect of lithium mining to fuel the battery boom of electric views of sustainability and also live on highly reflective land that is ideal for launching solar balloons.

What I am championing here is not this project or the technology it is imagining, but a motley group of people linking arms to attempt to reformulate our relationships with the extra-human environment in the long term. This is a rethinking of the world that is not just fabulating but storytelling through physics, social justice, material sciences, craft skills for prototyping, and social history and theory to the extent that it becomes a real possibility to be enacted.

We can drop back down from this borderless vision, aloft in the absolute quiet while traveling hundreds of miles an hour along the Gulf Stream, to more earthly modes of living differently. The science fiction of Octavia Butler, for example, has inspired multiple Black intentional communities that variously seek to "remember and reimagine our relationship to ourselves" (Earthseed, n.d.) and "focus on self and community healing" (Land Based Jawns, n.d.) in addition to food sovereignty (cf. White 2018). In a nation built on privatized stolen land and governed by illusory accountability systems, building and rearticulating relationships and experimenting in co-ownership beyond matrimony are all vital steps toward making alter-engineering possible and bucking homesickness. As one member of a cooperatively owned Black farm and cultural center in North Carolina has noted, "Building trust with one another. That feels like the work. Some of the most important work I'm doing in my life" (Purifoy 2020).

The mobile home, that has featured in this book largely as the ultimately sick home, even yields powerful opportunities for alter-engineering. The division of mobile homes from real estate ownership that has traditionally served as a source of structural insecurity is also an opening to escape the private property and predatory finance moorings of much of this continent's racialized homesickness. A countercurrent to the increased consolidation of mobile-home park ownership by hedge funds is the steady proliferation (and maintenance of) resident-owned manufactured-home communities, now numbering over one thousand parks across what is now known as the United States. These communities collectively own the land of their neighborhood and individually own their own manufactured homes. Such cooperatives open up pathways to collective landownership, regardless of racist credit history or discriminatory allocation of citizenship status. Some, such as the Dulce Lomita Mobile Home Cooperative in Asheville, North Carolina, allow residents to "make payments according to what it is that you make" (Barrows 2017) and have helped residents leapfrog from living in parking lots into collective ownership.

The people of Dulce Lomita see themselves as a community that is protecting a legacy that "low income people, and black people, and immigrants have been building for a long time" (Barrows 2017). The cooperative park itself has spawned an antigentrification nonprofit that has become a node in a national community wealth network of locally administered low-interest loans to channel "investment to marginalized communities that have faced the brunt of the extractive economy, deindustrialization, and systemic discrimination" (Seed Commons, n.d.). This mobile home community has, through the national network, moved over $6 million into worker-owned businesses and other resident-owned mobile-home parks to create permanently affordable housing in their neighborhood and to spread cooperative assets. The community practices collective, deliberative decision-making for public policy, zoning, and investment recommendations.

While the homes in this park were stripped to the studs and rebuilt, I worry about relative energy poverty as compared to conventionally built homes and, in other parks, about off-gassing materials. Most of these cooperative parks appear to be made up of homes on the older side, meaning the thermal issues may outweigh chemical exposure issues. I bring this up not to attempt to poke holes in their inspirational work but to underline how the methods advanced in this book to cultivate otherwise futures that overcome the conditions of racial capitalism can reproduce some of the issues that were the core subjects of this book (toxic homes). Just like holding together disparate practices and theories of change, holding together the method, condition, and subject of this book help to mitigate retrograde interventions. And it can further indicate where engineers and critical environmentalists might work further to materialize protective low-income housing (Grealy et al. 2022; Grealy 2023).[30]

If this section has you scratching your head, remember that even though chemical toxicity was the route into this extended meditation on homesickness, it is only one of three currents that I've outlined in this book and is inextricably caught up in the others. As a predatory home-loan system coconstituted the proliferation of formaldehyde-based engineered homes, it appears more than appropriate to understand community-operated financing as part of eroding the many buttresses that support a chemical that is currently too big to fail.

For further momentum away from the energy investments that maintain homesickness, those who already own a home could take the small step toward nonelectrical solar futures with the installation of a solar water heater (as opposed to solar panels). Following Indigenous directives for the return of public lands, 48 percent of all land in Western states, to Indigenous control and jurisdiction as a stepping stone to full #landback (Gilio-Whitaker 2022) is another ingredient in the not fully known antidote to homesickness. Back in the Gulf South, I also turn to the work of Monique Verdin (2019), whose words charge the first page of this book. Her Land Memory Bank and Seed Exchange catalogs the twin land losses and cultural losses of southeastern Louisiana, and it preserves the plant and cultural heritage of Native wetland communities. This preservation of memory, practice, story, and biological diversity then can feed into the radical change-making of the multiracial and multiethnic coalition of women of color that leads the Another Gulf Is Possible Collaborative, which she too co-leads, and seeks radical regional change from Brownsville, Texas, to Pensacola, Florida.

Arising from the semiarid grassland ecosystems of the North American Great Plains, the Land Institute is working to re-perennialize wheatgrass and other staple crops. Moving from an annual tillage-based agriculture to perennial polycultures, as the group seeks to do, has cross-cutting benefits spanning climate change, soil erosion and degradation, and biodiversity loss (Krug et al. 2022). This alter-engineering of both plant species through cross-pollination and selective breeding and of attendant agricultural practices and logistics can wean vast markets off of fossil fuel–derived fertilizers and pesticides while reducing water use and sequestering more carbon in the plants' elaborate root structures. Not far away, the Kansas City Tenants Union is scoring historic wins for tenants' rights and affordable housing, including winning four seats on the city's six-member district city council and ratifying a Tenants Bill of Rights.

From the prairies to Appalachians Against Pipelines, currently being pummeled in their fight to stop a fracked gas pipeline in West Virginia, to rural and small-town organizers in the Pacific Northwest, notably the Rural Organizing Project and Firelands, white people are playing important roles in explicitly and intentionally multiracial coalitions seeking to differently engineer the systems that maintain and advance homesickness.[31]

Admittedly, I sprain my imagination as I attempt to think through how accountability should and could be reengineered so that people like Betty, Mac, and Lizzie are not forced to let go of any hopes for responsibility or redress. Just like reconceiving one's relationship with energy, developing the institutions and relationships needed for more robust forms of accountability is first an imaginative project (Gilmore 2002; A. Davis 2003; Kaba 2021). All over the continent, community accountability initiatives have (re)emerged—predominantly through the stewardship of radical women of color and trans and Indigenous people (Tso 1992; Zion and Yazzie 1997; Nielsen 1999; Kaba and Hassan 2019; Peña 2019; Creative Interventions 2021). Often operating outside traditional civil and criminal courts, these processes center on ascertaining the needs of the harmed, and, in an uncanny echo of Tina's instigating question in chapter 4, determining who is obligated to address these needs (Dixon and Piepzna-Samarasinha 2020).

These grassroots or tribal practices are most readily associated with small communities and interpersonal wrongdoing. Understanding how these practices can provide value for communities in pursuit of accountability for environmental harms perpetrated by corporations and the state requires more work. Some of these accountability practices have undergirded Truth and Reconciliation processes that have adjudicated massive and distributed state violence in apartheid South Africa, Pinochet's Chile, settler colonial Canada, and Philadelphia after the MOVE bombing by police. Like the attempts at litigation in chapter 4, there were expansive problems with implementation (Madlingozi 2007; Clarke 2019). Those in power tend to be more interested in skipping to reconciliation when the aggrieved are just getting started in detailing the truth (Mazo and Pender-Cudlip 2018). To similar ends, the Christian ideology implicit in the South African Truth and Reconciliation Commission cast those who muster forgiveness as "good victims" in a way that ensured "previous material and social privileges are maintained" (Madlingozi 2007: 77–98). These past failings, among many others, help to clarify what justice looks like as demands for justice now include demands for better processes of accountability.

These have been just a handful of steps toward addressing homesickness and its immiseration. All are very particular steps that fit some issues, people, and places better than others. Not all of them are legible as alter-engineering topically—some are focused on the antipolitics of saying "no"—but their methods of organizing and community self-determination help realize worlds beyond racial capitalism. These are not plug-and-play fixes but commitments to experimenting with living through different values than those that structure dominant ways of building homes and happiness.

This book does not conclude with a recipe, but it does end with some guardrails as I turn to a partial set of questions to guide the development of engineered systems toward protection, equity, and accountability. Below are questions that I would ask myself about the projects (or policies) I might support and consider alter-engineering. This practice of advancing reflective questions builds on a long public rhetorical technique used by organizers such as Critical Resistance, Mariame Kaba, and Dean Spade (Kaba 2014; Spade 2016; CR 2021).

Questions that yield "no" to be deserving of time and energy:

- Does this project fetishize a chemical (or other downstream agent of harm) over the technosocial relations that underpin the scale, persistence, or inequities of exposure?
- Does this project shape-shift violences so that they fall outside of prevailing rubrics of harm (e.g., from greenhouse gases to mining and land-use issues, or moving trailers out of the region where they are understood as toxic)?
- Does the project set the problem too small?
- Is the imaginative horizon of this project limited by conservative ideas of feasibility (which may indicate where priorities and investments need to be shifted rather than being a truly impractical project)?
- (Of course) Does this project build something that will have to be undone later?

Questions that yield "yes" to be deserving of time and energy:

- Does the process of advancing this project reflect the values that ground it?
- Was this project conceived by or with the communities that face or have faced disproportionate negative impacts from the issues at hand?
- Is decision-making in this project governed equitably across stakeholders?
- Does this project seek to make "necessary" harms unnecessary rather than optimize them (i.e., not a project of making harms more efficient)?
- Does this project treat the extra-human environment not as a resource (a sink for human emissions, for example) to be managed but as a set of forces to be met?

This list cannot be finished, or approach being finished, by a single person. Additionally, there are many other guides to think through any alter-engineeringish project, from how to first get grounded in action when moved to fight injustice (Kaba 2018), to utilizing harm-reducing data practices (Cifor et al. 2019), to understanding the responsibilities of allyship (Gehl and Algonquin Anishinaabe-kwe 2012), and to using the processes of collaboration (Principles of Working Together Working Group 1991). Strike a line through the questions above that do not land for you. Or, if you like, fill the blank spaces, and more, with the questions that should be added.

My point in ending this book here is not to divert all energy to radical technosocial transformation and away from other approaches to change. That would be just as violent as the current world that renders such ideas as farcical. Doing so would overlook present suffering, inequity, and violence in the pursuit of an alter-engineered future. Keeping things as they are hands over the future to the dominant trajectories of environmental change and will likely never achieve escape velocity from core environmental problems. My point across these two final chapters in particular is to kindle the friction from these multiple logics, theories, and timelines of change into the warm glow of work that recognizes mutual value and mutual necessity without indulging in the desire to resolve differences.

As the FEMA trailers continue to decompose into the landscape, their metal chassis often all that persists, they leave us with much more than the stopgap shelter they provided and the harm that they caused. Exposures within these bare-bones homes ground this book in specificity and then fan out to innervate large issues across quests for more equitable worlds more generally. The death-dealing "solutions" of the FEMA trailers cast a long shadow. From the vile and unambiguously reprehensible coverup of mass harm, to paradoxes at the core of late industrial life, the continuum of injustices that manifest in their walls is instructive for conceptualizing harm and struggling to resist and move beyond it. As these superlatively simple homes have taken on the qualities of a house of mirrors over the course of this book, it is not always easy to look into the reflections they cast on even well-intentioned pursuits of change. But in the haunting images that fill many of the pages of this book, and the lurching—if not stumbling—pursuit of less toxic futures, are key lessons for detoxifying homes, concepts, subjectivities, and infrastructures.

ACKNOWLEDGMENTS

From my very first interview, I could already feel the conventions of single-authorship ethnography straining. After the last sentence was punctuated, the tradition feels to have fully buckled. This book relies on so much collective wisdom that my acknowledgments will surely be as insufficient as this book's single byline. Those whose observations and experiences fill this book are owed the ultimate gratitude. None of this work would have been possible without generous invitations into homes, workplaces, and fast-food joints. Thank you for trusting me with your stories and analyses. This book is dedicated to the people gripped by afflictions detailed here in such an acute way that they were never able to find me and will never know this book even exists.

A casual conversation with Dakota Moe and an open-minded invitation by Aaron Alquist drew me into the legal fray, which inspired the rest of this book. Becky Gillette was supportive and insightful from the start. Her tireless work on formaldehyde spans more than a decade and enabled every single word. Thanks also to Jimmy Bankston for teaching me about the intricacies of trailer construction and maintenance. To Harriett McFeely, Sharon Stinson, Charlie Syrie, Gina Phillips, Jessie Dilbeck, Frannie, Jennifer Jackson, Audrey Evans, Paul Stewart, Marty Horine, Ron Fowler, Nancy Martinez, and numerous others, thank you for your persistent insights and perseverance. All I can hope is that I have done right by you.

This book was written across twenty different apartments or houses, including in Lenapehoking, the traditional homelands of the Lenape; Tovaangar, land traditionally care-taken by the Gabrielino/Tongva peoples; Chahta Yakni, the traditional homelands of the Choctaw; as well as the traditional land of the Huron-Wendat, the Seneca, and the Mississaugas of the Credit. I recognize and acknowledge these traditional caretakers and am grateful for the opportunity to work on their lands.

Thanks to Elisabeth Hsu for encouraging me to take on this project as a master's student and to pursue doctoral studies. Thank you for teaching me what robust ethnographic writing means and for incisive comments. Javier Lezaun was nothing short of life-saving (even more so with the roping in of Ann Kelly who supercharges any idea). His calm redirections and encouragement recalibrated both my work and my understanding of what it means to be a generous scholar.

Kim Fortun, Mike Fortun, and the rest of the Asthma Files team laid the early foundation for this book and the collaborative form of my work. Kim inspired this project—both methodologically and theoretically—and facilitated and lent material support to the analytical chemistry aspect of this book, for which I remain thoroughly indebted. Brandon Costelloe-Kuehn was an indispensable partner throughout the formaldehyde testing substudy. I owe an enormous amount of gratitude to Marty Spartz and the rest of the staff at Prism Analytical Technologies for their consistent help and collaboration. Likewise, Linda Kincaid lent time, expertise, technology, and inspiration. When I give talks about Linda's work, audience members invariably respond with some variation of "she's so cool." I feel the same way.

The wisdom of friends and colleagues have improved this book enormously. I am especially grateful for comments from Jason Pine, Jerry Zee, Alex Blanchette Emilie Cloatre, Caroline Potter, Peter Kirby, Ali Feser, Biao Xiang, Natalie Porter, Tristan Jones, Rose Deller, Nerea Calvillo, Christabel Stirling, Ari Braverman, Ella Butler, Martyn Pickersgill, Nadine Levin, Noémi Tousignant, Christien Tompkins, Murphy, and the Oxford affect theory reading group. Thanks also to Lashon Daley for contributing a helpful detail. Elizabeth Povinelli was the dream external examiner and gave me crucial advice.

At Oxford, material support from the Philip Bagby Bequest of the Institute of Social and Cultural Anthropology, a fieldwork grant from Green Templeton College, and the University Vice Chancellor's Fund helped make ends meet. A National Endowment for the Humanities fellowship at University of Southern California provided a pivotal month of reflection and interdisciplinary exploration in the middle of my fieldwork. I feel truly grateful for that time with Tara McPherson, the Vectors team, and the other summer institute fellows. Two fellowships at the Science History Institute supported historical research and bestowed me with a supportive community of friends and colleagues. Jody Roberts was an incredible boss, shielding me from a tidal wave of bureaucratic nonsense and dangerous misinterpretations of my work by powerful people while also charging me to think more critically and systematically about toxics. Nasser Zakariya was a dream colleague, and both his and Jody's insights are deeply woven into my final chapter (thank you!).

In the process of researching and writing, I have leaned on my friends and want to thank those who lent camaraderie, much-needed distraction, and enduring support (Robert Rapport, Jonah Rimer, Naima Woods, Roger Norum, Sam Feather, Alice Ream, Paul Cavanagh, Alejandro Reig), company and patience on long road trips (Akasha Rabut, Katy Jane Tull, Veronica Hunsinger-Loe), and shelter when I was underfunded (Nikki Thanos, numerous couchsurfing.org hosts). I was lucky enough to have friends to collaborate with on this caper.

Akasha Rabut, thank you for enduring exposures with me in Louisiana, Oklahoma, and Texas and for your photos that animate this text. A huge amount of gratitude needs to be extended to Jakob Rosenzweig for spending days with me poring over hundreds of thousands of data points to fashion the resale map. Andrew Curtis and Michael Athanson also provided valuable mapping support. It was always a gift to discuss scientific apprehension and statistical eventfulness with Nicole Novak.

A whole world of experimenting with governmental accountability, the limits of the settler state, and the permission to pollute regulatory system was created with Environmental Data and Governance Initiative (EDGI). That world doesn't surface in this book, but the years of around-the-clock work on that project certainly invisibly edifies the conclusions of the final chapter. Big thanks to the OG EDGIers: Lindsay Dillon, Rebecca Lave, Gretchen Gehrke, Murphy, Lourdes Vera, Dawn Walker, Stephanie Knutson, Phil Brown, Sara Wylie, Eric Nost, and Chris Sellers.

The global community of Public Lab brought me into collaborations and taught me new skills that I could not have foreseen and helped me to articulate how research questions might not just indicate problems, but help address them. I'm indebted to Liz Barry, Gretchen Gehrke, Matthew Lippencott, Jeffery Warren, Shannon Dosemagen, Stevie Lewis, Chenier "Klie" Kliebert, Leslie Birch, and many others. Thanks to Christophe Guerin for ethically materializing our dream and Garance Malivel for cultivating fertile grounds for collaboration. This work at Public Lab pulled me into nearly a decade of collaboration with the always insightful and precise Karen Dannemiller and her many incredible students. Big thanks to Elizabeth Lara for vacuuming microbiome samples in Georgia and California and Gabby Resch for helping me 3D-model a key component. Thank you to Frank Finan, Rebecca Roter, and Joan Tibor for always patiently getting me up to speed.

The Aerocene community was lifesaving when I couldn't conceive of anything other than empirical trench warfare. Big shout-out to Tomás Saraceno, Sasha Engelmann, Bill McKenna, Derek McCormack, Kiel Moe, and the many others in that orbit who helped keep that part of the book aloft. The Technoscience Research Unit crew remains undefeated and I learned so much from simply being in the same room as y'all. A delightful hodgepodge of folks helped me think through various parts of the book in big ways and small: Alex Nading (still the best feedback I've ever received), Des Fitzgerald, Andy Balmer, Eben Kirksey, Joe Masco, David Bond, Simone Giertz, Amy Moran-Thomas, Joe Dumit, Juno Salazar Parreñas, the many alterlifers (esp. Vincanne Adams, Kelly Knight, Lochlann Jain), Max Liboiron, Murphy, Sophia Jaworski, Dominic Boyer, Cymene Howe, Ali Kenner, Gwen Ottinger, Emily Astra-Jean Simmonds, Etienne Benson, Jess Varner, Halle Young, the Swines, and loonch. Special thanks to Joe Masco for inspiring me to go to grad school and being even more wonderful once I finally met you.

At the University of California at Los Angeles, I was lucky enough to host a book workshop with Murphy, Hannah Landecker, Bharat Venkat, Chris Kelty, and Rafik Wahbi. Your insights carried this book into existence. Vincanne Adams has been teaching me for years about what a champion looks like. Bharat Venkat, Jessica Cattelino, Grace Hong, and Jessica Lynch provided important guidance and pastoral care. Two anonymous reviewers wrote the ultimate love letters to this book, not just spouting praise, although that was nice, but pointing out every single flaw and helping to envision a way to address them. The care and grace that went into those reviews is something I'll keep learning from long after this book. Thank you. Rachel Smith and Lilia Diaz (who called themselves my right and left hands: R + L) provided invaluable support in editing and citations together as well as clutch bibliography help from THE Abril Guanes. Regina Higgins provided deluxe copy and developmental counsel. Ken Wissoker was both deft in his guidance and extremely patient as I juggled entirely too much. Kate Mullen and Bird Williams provided extraordinary eagle eyes in the final stages. Ideas on Fire taught me so much in how they built the index.

The Skyscrape Foundation and the Max Planck Institute for the History of Science came through with some vital support during the final sprint when I was being pulled into a million directions other than finishing this book.

I cannot overstate how much I've been influenced by M Murphy and how their insights have buoyed the most important parts of this book. Brilliance is a concept hollowed out by academia but I vividly relearn what it should mean every time I'm in the same room as you. Max Liboiron tussled with the details of this book more than anyone. Thank you for running straight into the infuriating snags of previous versions and jujitsu-ing them into force-multiplying connections.

My ultimate thanks are reserved for my family—Michael, Lisa, Kate, and Neena—who taught me all I know about curiosity, thoughtfulness, working with passion, and trying to live right in a messy world. I'm sorry for always coming home in pieces, knowing that I would always leave whole—reassembled by your unwavering love.

APPENDIX I

FORMALDEHYDE INHALATION EXPOSURE STANDARDS

Organization	Exposure Limit	Additional Information
Environmental Protection Agency (EPA) Integrated Risk Information System (IRIS)	0.007 mg/m³ (~6 parts per billion (ppb))	Reference concentration (RfC) is an estimate of lifetime continuous inhalation that is likely to be without risk of deleterious noncancer effects during a lifetime. This particular RfC focuses on decreased pulmonary function, prevalence of current asthma or degree of asthma control, and allergic conditions. The same document that created this RfC also states that adult respiratory tract pathology can begin at half this concentration.
Agency for Toxic Substances and Disease Registry (ATSDR)	8 ppb (≥ 365 days)	Minimal risk levels (MRLs) estimate the daily level to which a substance may be exposed without the likelihood of adverse, noncancer health effects.

(Continued)

National Institute of Occupational Safety and Health (NIOSH)	16 ppb	Recommended exposure limits (RELs) are time-weighted average (TWA) concentrations for up to a ten-hour workday during a forty-hour workweek.
Federal Emergency Management Agency (FEMA)	16 ppb	This standard was specifically for FEMA trailers and has been withdrawn.
Green Building Council Leadership in Environment and Energy Design (LEED)	27 ppb	For a building to achieve LEED certification, indoor formaldehyde levels cannot exceed 27 ppb during clearance tests performed prior to building furnishing and occupation.
Agency for Toxic Substances and Disease Registry (ATSDR)	30 ppb (>14–364 days)	Minimal risk levels (MRLs) estimate the daily level to which a substance may be exposed without the likelihood of adverse, noncancer health effects.
World Health Organization (WHO)	81 ppb	This guideline was created to protect against sensory irritation in the general population for exposures of thirty minutes.
US Environmental Protection Agency (EPA)	100 ppb	Recommended maximum indoor air level. Exposures above this level can cause watery eyes, burning sensations in the eyes and throat, nausea, and difficulty in breathing.

APPENDIX 2

FEMA TRAILER TEST RESULTS (2011–2012)

City, State (# if multiple)	Type	Tester	Test Date	Formaldehyde Results in Parts per Billion (ppb)
Pasadena, TX (1)	Travel trailer	Shapiro	11/4/2011	162.60
Pasadena, TX (2)	Travel trailer	Shapiro	11/5/2011	63.41
Pasadena, TX (3)	Travel trailer	Shapiro	11/5/2011	105.69
Fultondale, AL	Mobile home	Shapiro	10/9/2011	113.82
Carriere, MS	Mobile home	Shapiro	10/13/2011	105.69
Oklahoma City, OK (1)	Travel trailer	Self-tested	2/25/2012	113.82
Hope, WV	Travel trailer	Self-tested	12/3/2011	113.82
Belleview, FL	Travel trailer	Shapiro	1/27/2012	121.95
Kentwood, IL	Travel trailer	Self-tested	5/16/2012	97.56
Kentwood, IL (2)	Travel trailer	Self-tested	5/17/2012	57.72
Oklahoma City, OK (2)	Travel trailer	Shapiro	3/28/2012	Chain of custody error
Denton, TX	Mobile home	Shapiro	4/1/2012	39.02
Tucson, AZ	Travel trailer	Self-tested	5/10/2012	34.15
Ralston, OK (1)	Travel trailer	Shapiro	3/31/2012	Chain of custody error
Ralston, OK (2)	Mobile home	Shapiro	3/31/2012	Chain of custody error
Mount Clemens, MI	Travel trailer	Self-tested	5/16/2012	50.41

(*Continued*)

Goldsboro, NC	Travel trailer	Self-tested	5/19/2012	47.15
Bremen, GA	Travel trailer	Self-tested	5/10/2012	138.21
Sandoval, IL	Mobile home	Self-tested	5/23/2012	69.91
Ray, ND	Mobile home	Self-tested	6/25/2012	43.09
Erath, LA	Park model	Self-tested	6/27/2012	22.76
Canton, MS	Mobile home	Self-tested	6/28/2012	38.21
Plant City, FL	Travel trailer	Self-tested	7/27/2012	69.11
Youngsville, LA	Travel trailer	Self-tested	8/4/2012	89.43
Town Creek, AL	Mobile home	Self-tested	8/13/2012	20.33
			AVERAGE	78.08

NOTES

1. Bulbancha, meaning "the place of many languages" in Choctaw, is the original name for what is now called New Orleans. Long before colonists set foot on the land, it hosted more than forty distinct Native groups. What is currently referred to as Lake Pontchartrain was originally known as Okwata, a Choctaw word meaning "wide water." The lands and waters of the Okwata basin have been inhabited by Indigenous peoples for upward of six thousand years and include unceded territories of the Choctaw, Bayougoula, Mougoulacha, Chitimacha, Oumas, Tangipahoa, Colapissa, and Quinipissa. See bulbanchaisstillaplace.org for more information.

2. The form of this discontinuous, albeit vital, mode of relating is similar to that identified by Zoë Wool in an American veterans' hospital, in which the specificities of harm "have a collectivizing effect, one that resides not in coherent narratives, testimonies or rallying cries, but in resonant common knowledge that lingers in bodies and the space they inhabit together" (2011: 3). The paradox of being simultaneously supported and worn down, isolated and brought together, is perhaps best synthesized by Benny Maygarden in this chapter's epigraph: "Formaldehyde is the only thing keeping me alive!"

3. The storm fell short of its predicted destructive force and focused its ire on less populated swaths of the region. In contrast to Katrina's lasting and multiple impacts, Rita's legacy primarily takes the form of evacuation policy (Litman 2006). The majority of deaths attributed to Hurricane Rita (90 of 108) were related to the massive and muddled evacuation process of between 2.5 to 3.7 million Texas and Louisiana residents (Zachria and Patel 2006).

4. The federal government has not always been responsible for providing postdisaster housing. The first instance was in 1969, when Housing and Urban Development (HUD) sent several thousand mobile homes to house those displaced by Hurricane Camille, which made landfall on the central Gulf Coast. This intervention was enabled by the Disaster Relief Act of 1969 (Moss 1999: 316) and the increasing federalization of disaster response due to Cold War civil defense security regimes (Rozario 2007: 156–73). The

provision of emergency housing to discrete disasters remains a federal responsibility today, now under FEMA.

5. Over 100 million cubic yards of debris were removed from the region, one hundred times more than Ground Zero in New York. Within a month of the hurricane's landfall, some 1.36 million diasporic Gulf Coast residents applied to FEMA for individual assistance grants. As displaced people fanned out to every one of the fifty states, hundreds of millions of pounds of emergency supplies were set in motion toward the Gulf Coast.

6. Very few services were provided to counter the isolation of these sites. Only one federally funded program attempted to mitigate the seclusion of group sites. The program, a bus service called LA Moves, was launched in January 2007, but by June of the same year it only served two of the hundreds of remaining group sites. The service did not provide transportation to welfare-to-work sites, employment, or human and medical services. Ridership was sparse (GAO 2008: 4).

7. Japanese American internment during World War II is the second-largest similar event (approximately 190,000 fewer inhabitants).

8. This process began when an emergency waiver of local rules against the placement of trailers outside of designated trailer parks expired on May 31, 2008. It wasn't until 2010 that the city council began planning how to enforce this regulation and remove the remaining 256 trailers (at peak usage, there were 23,000 FEMA trailers in New Orleans). The trailers were said to be "preventing recovery, lowering real-estate values for neighboring homes and serving as unwelcome reminders of struggles that residents want to move past" (Cohen 2010).

INTRODUCTION

1. This line of argumentation is made possible by an explosion of critical studies of toxicity in the early twenty-first century: Murphy 2006, 2017; Auyero and Swistun 2008; Frickel 2008; Roberts 2010, 2014; Guthman 2011; Landecker 2011, 2019; Agard-Jones 2012, 2013; Chen 2012; Boudia and Jas 2013, 2014; Cram 2016; Liboiron 2016, 2021; Nading 2017, 2020; Simmons 2017; Calvillo 2018, 2023; Davies 2018; Kenner 2018; Liboiron et al. 2018; Tironi 2018; Wylie 2018; Blanchette 2019; Hepler-Smith 2019; Thylstrup 2019; Balayannis and Garnett 2020; Feser 2020; Grandia 2020; Lefève et al. 2020; Rubaii 2020; Shadaan and Murphy 2020; Spackman 2020; Boudia et al. 2021; Müller 2021; Ntapanta 2021; Saxton 2021; Wakefield-Rann 2021; Bond 2022; Jaworski 2022; Adams 2023; Mah 2023; TenHoor and Varner 2023; Varner, forthcoming.

2. Deep gratitude to M Murphy for the succinct insights that breathed life into this paragraph.

3. These prompts are in line with Charles Hale's methodological push for anthropology to expand its commitments beyond primarily or exclusively serving "institutional space from which it emanates" (2006: 104) and extends that analysis to include methods far beyond the standard social scientific toolbox. As I tell my life science students in order to underline the importance of the social sciences and humanities, "Problems identified by life sciences often aren't remedied by life science approaches alone." The same is true of anthropologically identified problems.

4. If you are having difficulty imagining why current approaches are insufficient and what alternative approaches might be, the former will be addressed two paragraphs below and the final chapter details the latter.

5. To gesture to how long-standing and how widespread this sticking point between short-term survival and full-scale upheaval is, over a century earlier socialist-theorist Rosa Luxemburg made a point similar to those in the national antifracking movement when she decried that mitigating harms through labor organizing would amount to "the suppression of the abuses of capitalism instead of suppression of capitalism itself" (2008: 90).

6. Anticipating in these instances is not desiring for things to go wrong. Although there are moments for some when science, litigation, regulation, or activism feel like they have all run their unproductive course and hope can only be imagined through some sort of swift destruction. Of cases like this, I think of an email that Harriett, whom you'll meet in chapter 2, wrote to me when I sent her a draft of that chapter. She noted that food tasted like tin foil most of the time, she has chronic acute diarrhea, and her husband's left eye now waters so much that he walks around the house with one eye shut. She summarized their state of being: "So, that is where we are; deteriorating, broke and going downhill. . . . I don't know how this will end. . . . I keep hoping we will get wiped out with a tornado or something so we can start over . . . otherwise, there is no hope for us."

7. As elaborated in chapter 3, this argument is deeply inspired by Elizabeth Povinelli's observation that "making die, letting die and making live are organized within and through a specific imaginary of the event and eventfulness" (2011: 134). But the crux of my argument is that the issue is not just that slow, agentless exposures are less likely to inspire public and scientific concern, as are the ongoing subclinical illnesses that result from such exposures. Understanding the logics of that imaginary is extremely helpful for understanding distributions of abandonment, but to think that substantive change, rather than harm optimization, can be attained through the cultivation of eventfulness is, in my experience, more often than not a pernicious trap.

8. Spaces of sanctioned adversarial contention are often designed to stabilize power and provide for the performance of accountability (see chapter 4). Of course, against all odds victories are possible, yet the terms of success are often deeply circumscribed. Fighting these available battles is important, but they are not the end of the road.

9. Data that I acquired by way of Freedom of Information Act requests suggest that, in the wake of the hurricanes, many more units were acquisitioned than inhabited, which is why this number is thirty thousand units higher than the number of units deployed.

10. As Laura Pulido (2016: 2) has touched on in the case of the nearly 40 percent of Flint, Michigan, residents who identify as white, white people can become devalued through their proximity to Blackness by living in a Black majority city.

11. The one exception to this depreciation trend is when mobile homes are placed on foundations on owned land and federal regulators categorize the trailers as real estate rather than personal property (Wilkinson 2020: 57). This scenario accounts for the minority of mobile-home tenure (19 percent in 2019) (Park 2022).

12. As noted in the US Commissioner of Indian Affairs annual report of 1877, Native Nations "were induced by the authorized agents of the Government to give up their old homes by the promise of assistance in building new ones. Yet I am informed that no provision has been made by the Government for building them houses or even assisting them to tools, nails, lumber, &c." ("Report of the Commissioner of Indian Affairs," 565–66).

13. To respect tribal sovereignty, no work conducted on tribal land is documented in this book. All work conducted on sovereign tribal lands was conducted as a pro bono service to individuals or tribal governments upon their request and is not sharable. I have, however, included research conducted off tribal land with consenting Indigenous interlocutors in this book.

14. The Recreational Vehicle Industry Association was worried that the glut of trailers would drive down the secondhand mobile-home market. After the court order to not sell the FEMA trailers expired, the industry attempted to lobby Congress and court aid organizations in the hopes of sending over 150,000 bargain-priced FEMA trailers to Haiti, which experienced a 7.0 earthquake in early 2010. This plan did not materialize, but the Clinton Foundation did send manufactured school buildings from Clayton Homes to Haiti. These were found to have elevated formaldehyde levels that made students ill (I. Macdonald 2011).

15. The manufactured housing market experienced a securitization-driven market meltdown in the late 1990s that foreshadowed the bursting of the larger housing bubble around a decade later. The boom in risky chattel loans was made possible by large increases in the securitization of manufactured home loans—from $184 million in 1987 to $15 billion in 1999. When the national housing-market bubble popped in 2007–8, hundreds of billions of federal dollars were dedicated to rescuing mortgages from default. This program did not include chattel mobile-home loans (Burkhart 2010).

16. Prism Analytical Technologies has since been purchased by Enthalpy Analytical.

17. The homogeneity control protocol for our study was as follows: The indoor air temperature would be maintained at 72–75 degrees for at least twenty-four hours prior to testing. For tests administered by homeowners, photographic evidence of the thermostat was required (clear camera-phone pictures at the time of testing were sufficient; preferably picture-messaged to the study telephone at the time of testing to verify the time of test). All windows had to remain closed for twenty-four hours prior to the test. All tests would occur at the same time of day. One-hour sampling began between 1:30 p.m. and 3 p.m., preferably at 2 p.m. Tests did not occur on rainy days or on days that were abnormally warm or cool, which was defined by local norms. Tests did not occur on windy days (defined by leaves and small twigs in constant motion, or wind that can extend a lightweight flag—level 3 on the Beaufort scale). The sampling device was placed on a nonformaldehyde-emitting surface in the middle of the bed in the master bedroom. Photographic evidence of the device location was also required.

18. Chapter 5 details the further multicontinental work involved in building open-source monitoring and remediation equipment that led to dozens of further tests.

CHAPTER 1. AT HOME IN THE SURREAL

1. His blog can be found at https://femarailer.blogspot.com/.

2. Prop 65 refers to a ballot initiative (a policy that is enacted by popular vote rather than through the legislature) from the State of California that enacted the Safe Drinking Water and Toxic Enforcement Act of 1986. As a result, items sold in California that contained chemicals that caused cancer or birth defects required a warning and, because of the size of California's market, products sold across the country are labeled.

3. By searching sales records for the vehicle identification number (VIN) of Meredith's trailer, I was able to find that Wade paid $4,060 for it on February 1, 2007, yielding a $1,975 profit if his storage costs were zero and he towed the units himself from the relatively nearby auction location and not including fuel. This is a similar profit margin to what he described in an interview as his standard profit margin. Meredith paid $1,500 to have the unit towed to Missouri.

4. These numbers derive from spreadsheets provided to me by the GSA in response to several Freedom of Information Act requests—the law that regulates public records release for federal agencies.

CHAPTER 2. CHEMICAL FETISHES

1. This was the only time my line of questioning provoked a critical appraisal of the smell. She had bought a FEMA trailer to live in while she saved money to move back home to the Southwest. She paused to think after I asked her why she liked the "new" smell. She thought out loud, "I guess you are supposed to think it's nice, but it's just chemicals from the materials used to make it." There can also be too much of a good thing, as one Michigander noted: "It's like the new-car smell times ten."

2. Phenomenological studies on pollution, environment, and well-being bring into crisp relief the intimate place-making and place-disrupting capacity of smells. They highlight how we often take displeasing scents as the primary indicators of environmental contamination (Fletcher 2005; Auyero and Swistun 2008; Brant 2008; D. Jackson 2011; Reno 2011). Since the article version of this chapter established its main line of argumentation (Shapiro 2015), a number of incredible analyses of atmospheric relations to bodies and politics have been published (Kenner 2018; Tironi 2018; Calvillo and Garnett 2019; Engelmann and McCormack 2021; Calvillo 2023).

3. See Grabbs et al. 2000 for a list of more than sixty chemicals identified in four different new vehicles. The total VOC level of one of the cars was two orders of magnitude higher than the outdoor air.

4. *Popular Mechanics* (August): 7.

5. *Popular Mechanics*, 1945 (August): 7.

6. *LIFE*, 1946 (June 3): 17.

7. *Dow Diamond*, 1960 (March): 1–2.

8. One example among many: In April and May 2007 the advertisement inserted into the bottom of Yahoo! Emails read, "Ahhh . . . imagining that irresistible 'new car' smell? Check out new cars at Yahoo! Autos."

9. The narrator of Ford Spain's web advertisement describes the product as follows: "We knew that not everybody could afford a new car but that was no reason to miss out on the excitement. To try to regain that feeling with an ad was hopeless. Our cars were already nearly new . . . but there was still something missing. So we created a product, a sensation."

10. Thanks to Pamela Dalton for helping to refine this point.

11. As a corollary, luxury cars tend to infuse their interiors with new leather as opposed to a new-car smell.

12. See Shapiro 2015 for an extended analysis of how this version of the sublime differs from older notions.

13. Within toxic domestic ecologies, the absence of mice and cockroaches or the frenetic behavior of companion birds, cats, and dogs are often read as sentinel indicators of toxicity (see *Limn*, no. 3, http://limn.it/issue/03).

14. She also used UMEx formaldehyde badges to validate her real-time results, as the copresence of other chemicals can increase or decrease the readout of the real-time meter.

15. The sleep disorders that Linda has recorded are not restricted to nightmares. She notes that several of her married clients now sleep in separate beds because of either insomnia or flailing around at night; both issues were reported to me by one in three of my informants.

16. For instance, the onset of the effects she claims formaldehyde has on her body (50 ppb) is 6 ppb above the California Office of Environmental Health Hazard Assessment's (COEHHA 2008) acute Reference Exposure Level for formaldehyde sensory irritation (44 ppb).

17. So-called green homes do pose a potential trade-off between energy efficiency and health. Such homes are sealed more tightly but often use the same construction materials as traditional homes, and as a result they are likely to bear high ambient chemical concentrations (Kincaid and Offermann 2010).

18. This phrase appeared in a letter from Betsy Natz, the executive director of the Formaldehyde Council, to Linda Kincaid, dated August 11, 2009.

19. Her husband, Dick, suffered from eye irritation, a loss of his sense of smell and taste, and numerous other symptoms shared with Harriett.

20. Her radon levels were negligible. A state employee checked for sewer gas and black mold. The couple's water well has cleared annual inspections from the state for nitrates and *E. coli* (as mandated as part of the license for Harriett's wedding catering business, which operates out of their home).

21. In Bowser's autopsy, dated April 4, 2008, J. N. Henningson, DMV, wrote in his comment: "Extramedulary hematopoiesis observed in spleen should not be seen in this age of dog. Regenerative nodules in the liver are common in all dogs. The hemorrhagic pneumonia is severe. . . . The hemorrhage in liver, lung and brain most likely led to the demise of the animal."

22. The appearance of zombie narratives, first encountered in Haitian folklore as parables of the woes of industrial capitalistic and colonial exploitation, can generally be read as indicators of environmental, economic, and scientific anxiety (Dendle 2007).

As the novelist I spoke with in Oklahoma is of Cherokee descent, the resonances of the figure of the zombie as colonial critique are especially poignant.

23. This political feeling "aggregate[s] life diagonal to hegemonic ways of life" (Povinelli 2011: 30).

CHAPTER 3. UN-KNOWING EXPOSURE

1. Email sent from Dan Shea to Stephen Miller on March 17, 2006.

2. She chose the location of her new home because of its low likelihood of experiencing another natural disaster.

3. It is not social happenstance that white, white-collar individuals with established ties to national organizations would be the ones to fund testing and circulate findings.

4. This chapter combines multiple approaches. Sociologist Linsey McGoey (2007: 217) has noted that strategic deployments of ignorance are supplanting plain-old state and corporate secrecy as dominant modes of information management, and science studies scholars have studied how a lack of knowledge is produced (Proctor and Schiebinger 2008). Strategic ignorance, the manufacturing of doubt (Michaels 2008; Oreskes and Conway 2011; Richter et al. 2018), and the subjugation of knowledge to the point that they "have been disqualified as nonconceptual knowledges, as insufficiently elaborated knowledges" (Foucault 2003: 7), are all at play in un-knowing. But not only. The unknowing in this chapter is also part of larger "regimes of imperceptibility" (Murphy 2006) that make sense of the patterned outcomes of scientific explanatory systems that time and time again found low-dose chemical exposures to be inconclusively toxic, undeserving of redress, and without an onus for remediation. And even beyond those technical horizons, as will, Un-Knowing Exposure be elaborated in the chapter, standard cheap tricks of delaying assessment, exercising secrecy, and manipulating expertise in addition to labyrinthine bureaucracy are at play.

5. Internal email sent from Patrick Preston to Jill Igert, Jordan Fried, and Kevin Souza on June 15, 2006.

6. The interagency relationships between ATSDR, FEMA, and CDC had solidified in the wake of 9/11 when FEMA "needed the presence of a medical-type person in a white jacket to give [community members] the assurance that everything was kosher, so to speak," as one ATSDR scientist described it to me. After that collective response, meetings with FEMA attorneys were not unusual.

7. Joseph Little, handwritten notes, "Formaldehyde Consultation," January 24, 2007.

8. Internal email sent from Patrick Preston to Scott Wright on November 30, 2006.

9. Renda had communicated this concern as early as the spring of 2006 when FEMA had asked him to vet a fact sheet about formaldehyde that they were releasing. He pointed out that the fact sheet did not list any chronic exposure risks.

10. The press release is available at http://www.fema.gov/news/newsrelease.fema?id =36010 (release number HQ-07-061), issued on May 4, 2007.

11. Backlash against scientists like Renda who questioned institutional pressure to dismiss chemical exposure concerns are not limited to ATSDR. For further examples, see Freudenberg 1984; and Brown 1992: 275. The bureaucracy had turned against Renda. He

would routinely return from lunch to find notes on his door asking him where he had been—someone was attempting to establish a paper trail of his poor work ethic. When his wife was getting bypass surgery, he remembers receiving a call challenging his use of family medical leave. His office was constantly reassigned.

Howard Frumkin, who would later disparage the "level of concern" in front of Congress, was a key figure in punishing Renda for voicing concern over the agency's analysis. In January 2010, a year after his role in corrupting the science of this report became known, Frumkin was reassigned from directing all of ATSDR to directing the CDC's Climate Change and Public Health program, a shift that cut his budget responsibilities by more than 97 percent and reduced the number of full-time employees under his direction by 99 percent (Sapien 2010). Frumkin has gone on to become a global leader on climate change and public health leading the Our Plant, Our Health initiative at the Wellcome Trust (the United Kingdom's largest charity) and a dean at the University of Washington School of Public Health. Frumkin and Tom Sinks, his deputy, were the two people who handed Renda his reassignment documents in Italy. As the Majority Staff Analysis of the Subcommittee on Investigations and Oversight in the Committee on Science and Technology in the US House of Representatives found, "Dr. Frumkin and Dr. Sinks' responses have been intent on confusing and mischaracterizing the events that unfolded regarding their own involvement in the formaldehyde issue, not only in interviews with Subcommittee staff, but during sworn Congressional testimony as well" (Majority Staff Analysis 2008: 36).

12. As Sara Ahmed concisely notes of raising dissent in institutional settings, "when you expose a problem, you pose a problem" (2017: 37).

13. When researchers adjusted for variables like relative humidity, temperature, and ventilation, they found average formaldehyde levels in the trailers produced by leading manufacturers to be over the 100-ppb level: Gulf Stream (111), Pilgrim (129), Forest River (102), and Keystone (122).

14. Larger pedagogical issues may feed into Smith's lack of clinical experience with formaldehyde-related asthma issues. As toxicologist Francesca Daniels, PhD, DABT (introduced in the following chapter), asserted to me in an interview, "Most MDs are not trained to recognize the environmental or occupational etiology. . . . Once you get outside some of these classic older etiologies of environmental exposures and occupational diseases, they're lost. They just haven't been trained. And they are not trained to do an environmental history or an occupational history."

15. This coordination is akin to the practice of "ontological choreography" that sociologist of science Charis Thompson (formerly Cussins) has identified, forging "a functional zone of compatibility that maintain[ed] referential power between" his different medical existences (Cussins 1996: 600).

16. My thinking on this is informed by Astrid Schrader's work on waterborne single-celled microorganisms that were simultaneously the suspected and vindicated culprit of large-scale fish die-offs, where she developed the notion of "phantomatic ontology." While the term was coined to describe scientific objects that "reshape their configurations in different contexts" (Schrader 2010: 278), it helped me analyze how the producers, or at least brokers, of scientific knowledge's own expertise bear "the paradoxical exis-

tence of a specter as neither being nor non-being" (278). As Lezaun and Woolgar (2013: 321–23) note, ontology and epistemology are not easily parsed. Both are implicated and entangled within broader practices of un-knowing links between formaldehyde and asthma.

17. The three mechanisms explained: first, IgE is immunoglobulin E and is believed to be a key agent in allergic and hypersensitivity responses. After initial exposure, formaldehyde-specific IgE antibodies are produced and circulated in the bloodstream. Due to elevated levels of corresponding IgE, the subsequent reentry of formaldehyde into the body can lead to an aggressive and rapid immunological response, yielding a large histamine release and resulting in inflammation and bronchoconstriction (Wantke et al. 1996). Second, the trigeminal-vagal reflex relates to nerves that weave into the mucus membrane of the nasal passageway (the trigeminal) and innervate the lungs. When an irritant is detected in the upper airway, a signal is sent through the brain to the lungs to decrease the respiratory rate and contract the smooth muscle surrounding the bronchial tubes. Third, the oxide synthase mechanism refers to the enzymatic connection between formaldehyde and asthma. The primary molecular defense against formaldehyde is an enzyme that bears receptor sites identical to those of the enzyme responsible for regulating an important bronchodilator. This means that formaldehyde binds both to the enzyme that destroys it and to an enzyme that is responsible for keeping the bronchial passageways open (Henderson and Gaston 2005).

18. Letter from Katherine Norris (CDC/ATSDR FOIA officer) to Nicholas Shapiro, March 5, 2015, #15-00400-FOIA.

CHAPTER 4. ENVIRONMENTAL LITIGATION

"Healing Without Accountability" section from B. Brown et al. 2020: 223.

1. When residents called a federal environmental hotline to see what steps could be taken to mitigate their exposures or to inquire about their consumer rights, they were simply told, "Get out."

2. The allure of thinking that the relationship holds in reverse, that the rubric for giving meaning to their abandonment is also a map for locating justice, is profound, and this is particularly acute in places of slow and chronic death.

3. Four years after Katrina, myocardial infarctions in New Orleans were three times more prevalent than before the storm (Hameed et al. 2011; Moscona et al. 2012). The same studies also noted statistically significant increases in unemployment, lack of medical insurance, medical noncompliance, substance abuse, psychiatric comorbidities, history of coronary artery disease, and percutaneous coronary interventions. There were no significant differences between the demographics of the two compared populations apart from an increase in those labeled as single or as divorced. While living in a FEMA trailer was not a noted variable in these studies, they are likely another stressor—just one among many (Adams et al. 2009; Adams 2013).

4. In an interview, a plaintiff expert witness, who asked not to be identified by his scientific specialty and only sparingly quoted, averred that a "close coordination" between scientist and lawyer was the key to winning cases. He felt that the acquittal of the defendants

in this case was largely due to an "abysmal" job of the lawyer-scientist coordination on the side of the plaintiffs.

5. The claim against the installers has to do with how the trailers' frames were torqued as each corner was jacked up individually before being placed on cinder blocks— transforming this temporary housing unit into a more permanent dwelling. The twisting of the frame broke seals and moisture barriers, allowing for water entry, which exacerbated formaldehyde off-gassing.

6. In this case, the court allocated 28.5 percent of the settlement funds from both the manufacturer (i.e., $10,678,543.64) and contractor (i.e., $1,461,836.26) classes to the consortium of attorneys. The plaintiff steering committee could not agree on how to divide these funds among themselves, so the court appointed a special master to help govern the process. The special master was paid $350 an hour out of the attorneys' pot of money. Nineteen firms were awarded percentages of the settlement, ranging from 0.46 percent to 27.10 percent of the fee pool.

7. To show the other side of the coin, the following is an email from an anonymous attorney, defending their role after a paltry settlement was reached in September 2012. This message was forwarded by Becky Gillette to her "formaldehyde list" on October 1, 2012: "I and the other lawyers in this case spent thousands of hours and millions of dollars in this case. I don't anticipate that I will make a cent and my firm will probably not be reimbursed for all of its expenses. . . . [The] vast majority of people simply felt that it was bad that people got sick, but that it just didn't rise to 'a federal case' since people received these for free. Many said these trailers were pieces of junk but not necessarily 'unreasonably dangerous.' Also, most plaintiffs smoked or had other health problems unrelated to their time in the trailer." These claims were not substantiated in the email.

8. A judge later overruled the multimillion-dollar punitive damages, and the plaintiff was awarded around half a million dollars for the burns that had covered 16 percent of her body and hospitalized her for a week. For more on this case, see the documentary *Hot Coffee*.

9. As an indicator of the exaggerated nature of reward for those injured, of the 25,000 tort and contract trials that reached conclusion in state courts in 2005, fewer than 3 percent (seven hundred) were awarded punitive damages. The median award amount in cases with punitive awards was $64,000 (Cohen and Harbacek 2011: 1).

10. This attorney was referring to the, at the time, ongoing FEMA trailer lawsuit and a lawsuit that he and a consortium of lawyers were considering filing on behalf of resold FEMA trailer residents. He was not saying that plaintiffs would receive no recompense at all but that the lawyers would be disproportionately paid because of the sheer number of plaintiffs.

11. I don't know whether Lizzie collected settlement funds for Rhett's injuries because she unfriended me on social media shortly after Rhett passed. I interpreted that as a desire to move on and did not bother her for more information.

12. Those impacted by perhaps the most widely known environmental justice issue in the contemporary United States, the water-quality crisis in Flint, have written about the difficulties generated by the proposed settlement from the State of Michigan: $600 million spread across at least 33,000 plaintiffs. The settlement was some seventy-two

times larger per person than the FEMA trailer litigation. Bob Brown and coauthors note that the processes for distributing funds were poised to become "sources of community division as plaintiffs compete over monetary awards, undermining the potential for collective action in the future" and that it was "unclear whether the settlement will enable undocumented community members access to the compensation they, too, deserve." They write that the settlement was not a "significant step toward justice" and that it "reflects a paternalistic pattern of thinking and decision-making evident in the crisis itself, in the official response to the crisis, and in underlying causes of the crises, such as white supremacy, systemic racism, and the orchestrated denial of local democracy" (Brown et al. 2020: 222).

CHAPTER 5. WORKING THE STOPGAP

1. This space center is named after former Mississippi Senator John C. Stennis, whom retired Lieutenant Commander Reuben Keith Green (2020) describes as "the heart, soul, and brains of the white supremacist caucus in the 1948 Congress." Use of evidence obtained by torture was banned by the Supreme Court (*Brown v. Mississippi*: 297 U.S. 278 [1936]) after Stennis, then a prosecutor, sought to execute three Black men for murder confessions that were extracted via torture. As a signatory of the Southern Manifesto of 1956, he was an ardent supporter of the indefinite continuation of segregationist Jim Crow laws. The naming of this space center after an infamous racist is indicative of bloody Mississippi history; even leaving Earth's orbit does not allow a space crew to escape from the legacies of institutionalized racism in what is now known as the United States.

2. This finding propelled NASA to develop strict off-gassing protocols for assessing materials brought into space. A colleague in the space industry in India noted with frustration that the expense of these exposure chamber tests is a key financial barrier for her company to compete with the big aerospace companies; in turn, that helps maintain US dominance of the market.

3. NASA results had been announced to great fanfare at the National Press Club in Washington, DC, in September 1989 (the first preliminary findings had been published in 1984). The EPA had been repeatedly asked to weigh in over the intervening years. Biosphere 2 had opened on September 26, 1991, and its first hippie modernist mission was underway at this time. Biosphere 2 was built to be a vivarium, an artificial closed ecological system that spans 3.14 acres, dwarfing the Biohome materially and historically.

4. The difference between the 2011 reading and the 2015 initial reading was 81.69 ppb, meaning that if the decay rate was constant (a big assumption), it would have been reducing by about 31 percent each year. Our observed 50 percent reduction over the course of a month would then account for ~1.5 years of reductions at that rate.

5. See Moran-Thomas 2017 for a biomedical corollary.

6. This system had been subsidized by the use of a $1,000 flow meter to gauge the flow of the vacuum. We substituted this highly expensive flow meter for a suitable and much cheaper flow-meter design that ranged from $15 to $75.

7. Thanks to Sara Wylie for introducing us to RI DOH.

8. Email sent from LumberLiquidators@e.lumberliquidators.com to a homeowner in July 2015. It is unclear how the test purports to understand the formaldehyde contributions of the floor compared to other sources or what the company's threshold for abnormal exposure is.

9. Because I conveyed my professional opinion to that homeowner that her results were very high, that likely also motivated the DOH into working with her to pinpoint the sources of her multiple VOCs (in addition to them being vexed by us).

10. I had seen this over and over again. Crises spur technology development that remains perpetually prototypical. For example, a VOC-sensing system I had helped to develop as a postdoc ended up having deep technical issues and essentially produced little to no useful data. See https://citizensense.net/frackbox-data/.

11. National Science Foundation Award #1645090. For this project, learning from our mistakes, we paid for expensive chamber testing and validation, including cross-contaminants like acetaldehyde and a VOC mixture (Zhang et al. 2019).

12. The two grants are Housing and Urban Development Award #OHHHU0057-20 and #OHHHU0078-24.

13. Special thanks to Liz Barry for sharpening this point.

14. Attempting to buttress the ethical stance of one's research by opening up another theory of change also brings with it a steep learning curve. The act of building a new sensor technology is both decentering (ridding concerned residents of the need for me to come to their house) and deeply arrogant (in thinking that a scrappy team can outperform the tens to hundreds of millions of dollars that have been spent in the development of existing tools).

CHAPTER 6. ALTER-ENGINEERED WORLDS

1. The 2017 omnibus appropriations bill called out this formaldehyde assessment by name, insisting on further scrutiny. Environmental Protection Agency (EPA) officials later noted the way that political appointees within the EPA suppressed the release of this assessment and further accused political appointees of leaking the draft assessment methodology to the American Chemistry Council, the primary chemical industry lobby, and fostering preemptive rebuttals of the not-yet-released risk assessment such as this one (Lerner 2021).

2. Corn silage is a fermented animal feed that emits a broad range of volatile organic chemicals (VOCs). Outdoor ambient formaldehyde, which is the top hazardous air pollutant (HAP) contributor to cancer risk in the United States (Strum and Scheffe 2016), stems not only from direct emissions of the chemical but also from a host of other anthropogenic VOCs breaking down through oxidation into formaldehyde as well as biogenic isoprene. Formaldehyde is responsible for over 50 percent of the total HAP-related cancer risks in the United States (Strum and Scheffe 2016). The list of the 188 compounds designated by the EPA to be HAPs is in "Initial List of Hazardous Air Pollutants with Modifications," www.epa.gov/haps/initial-list-hazardous-air-pollutants-modifications.

3. This introductory section was first drafted in 2017. The ACC-funded article is Mundt et al. 2017. The formaldehyde poster is Zhu et al. 2017. As the following paragraph indicates, the day that I wrote this section was indistinguishable from the day that I revised it in 2024.

4. As investigative reporter Sharon Lerner (2021) and Lerner and Shaw (2024) have detailed, industry lobbying has intensified since 2020 as health impact assessments inch closer to fruition. In the four years prior to publication of this book at least seventy-five trade groups have attempted to suppress or disqualify the health impacts that the EPA might ascribe to formaldehyde. These include: the Fertilizer Institute, the Golf Course Superintendents Association of America, the Toy Association, the National Chicken Council, the Asphalt Roofing Manufacturers Association, the Independent Lubricant Manufacturers Association, the RV Industry Association, the Halogenated Solvents Industry Alliance, and the ACC .

5. The petition included support from the Sierra Club, twenty-five other organizations, and approximately five thousand individuals. Letter from Tom Neltner, Sierra Club, to Stephen Johnson, Administrator, Environmental Protection Agency, Re: Citizen Petition to EPA Regarding Formaldehyde in Wood Products, March 20, 2008.

6. Biogas of this sort will only ever contribute very small amounts of methane, but its mere existence serves as an alibi for continued natural gas extraction and usage by creating the idea that potentially some natural gas used in homes might not be extracted through drilling (Saadat et al. 2020).

7. Benefits from outdoor air regulation have been highly uneven in raced and classed ways (Nunez et al. 2024).

8. Clay was the director of environmental and regulatory affairs at Koch Industries, which owns Georgia-Pacific, a top formaldehyde producer. While employed by Koch he sat on the EPA Clean Air Act Advisory Committee for more than a decade (1998–2009). At his time of death in 2018, Clay's son and daughter-in-law were both attorneys for EPA region 6.

9. Gorsuch and some twenty EPA officials were pressured to resign in 1983. Clay was not one of them, but the "dubious formaldehyde risk analysis" was one of a number of controversies that led to their ousting. See Powell 1999: 30.

10. In 1987, OSHA set an eight-hour permissible exposure limit of 75 parts per billion (ppb) for offices, schools, factories, and other indoor workplaces. In 1985, HUD began regulating formaldehyde emissions from the composite woods utilized in manufactured housing. It set the upper limit for plywood at 200 ppb and for particleboard at 300 ppb. These commodity-based regulations were designed to yield an interior atmosphere that contained less than 400 ppb (see CSC 1997: 4). For comparison, the World Health Organization's maximum recommended limit for a mere half-hour exposure, not the day-in-and-day-out exposure incurred in the home, is more than four times lower (81 ppb) than this intended but unmonitored threshold. The Agency for Toxic Substances Disease Registry (ATSDR) notes that chronic formaldehyde exposure over 8 ppb can cause nega-tive health impacts in humans, a level that is fifty times lower than the intended air quality furnished by the HUD regulations. Further, enforcement of HUD's standard is lacking, and the standard is openly disparaged by other governmental agencies as "essentially volun-tary" (see California Air Resources Board 2007).

11. The only domestic US air-quality regulation (as opposed to building material regulations) was announced by FEMA in April 2008 and exclusively applied to their mobile emergency housing units. It's unclear if they quietly rescinded their maximum allowable level of 0.016 ppb in October 2011 because the media attention after Katrina had abated or due to the sheer difficulty of meeting this standard. While only short-lived, this shooting star of regulation indicates what could have been some thirty-five years earlier.

12. The company further sought influence through sitting on an advisory panel for the National Association for the Advancement of Colored People (NAACP), attempting to leverage the environmental justice angle to further their products. Interestingly, this published study notes no conflicts of interest.

13. Shorting stocks is a way that speculative investors "bet against" the market, essentially making money when particular stocks go down.

14. In 2018, LL paid $36 million to settle a class-action lawsuit with the 760,000 customers who bought LL's laminate flooring between 2009 and 2015 (*60 Minutes* 2019).

15. Thanks to Evan Hepler-Smith for pointing me to this article.

16. Indicating the foundational importance of the chemical, formaldehyde is the first entry in this chemical classification system that contains over 204 million unique chemicals.

17. Because of the increased weight of electric vehicles from hefty batteries, they require more roadway maintenance (i.e., bitumen). They maintain the same ten million miles of ecosystem-segmenting, hydrology-altering, ossified oil spills that are roadways. Resuspension of particles and tire and road abrasion are intrinsic to automobility. One modeling study noted the minuscule air-quality impacts of electric vehicles (EVs): "Fleet electrification has a limited impact on PM10 reductions in terms of both emissions (3–4%) and air quality (2–5%)" (Soret et al. 2014: 62). One could also think of what social and material relations of hydrocarbon worlds are maintained in a world fueled by solar panels. Think of the quartz mines, the highly energy-intensive (and very often highly toxic) process to produce solar-grade silicon, the 6.9 million gallons of water used annually to clean utility-scale solar installations (or the 400 million gallons used during construction) (Mulvaney 2014), the sensitive desert ecosystems that are increasingly wastelanded (Voyles 2015) by large arrays, and the operational use of toxicants in dust suppressants, rust inhibitors, antifreeze agents, and herbicides (Hernandez et al. 2014). Think of the demand for cobalt that would quadruple if the world's electricity and transportation infrastructures ran entirely on renewables, most of which is mined in the Democratic Republic of the Congo and implicated in child labor, human rights abuses, and poor worker rights and conditions (Dominish et al. 2019).

18. They now distinguish not between reformist reforms and nonreformist reforms but between reformist reforms and abolitionist steps (Critical Resistance 2021).

19. In outlining their "deal," the authors are clear that the Green New Deal (GND) that aims to transition to "clean energy" by 2030 and furnish sixteen million jobs in the process is insufficiently intersectional. This is in part due to the racialized labor discrimination of mass incarceration (see Western and Beckett 1999) and the large scale of undocumented migrants, yielding *half* of the workforce as potentially ineligible to benefit

from GND jobs due to multiple forms of criminalization (Red Nation 2021: 29). The authors advocate for divesting from the military, policing, and incarceration for both harm reduction and funding expansive changes.

20. Indeed, US denunciations of this constitution often pivoted on concerns about nationalizing Chile's mining industry, which would complicate lithium supplies needed for fleet electrification. In this case, the way different homesicknesses reinforce each other is made tragically clear.

21. See, for example, https://millionexperiments.com/.

22. Okay, maybe not the anthropocene. This is the initial and most limited -ocene that has increasingly been captured by a narrow focus on a mid-twentieth-century start date for this epoch (cf. Ellis 2023).

23. These questions arose from conversations with Jody Roberts and Nasser Zakariya. The years I spent working with them both were so vital for knocking me off my instrumentalist trajectory and into a more critical posture. I am perpetually grateful.

24. I first came to this project by way of a combination of my professional and social networks and being available enough to answer "yes" to a last-minute invitation, knowing almost nothing of the project. My introduction to this work by way of a defined problem is readable but not necessarily true to how I got involved. Sometimes one encounters a path to an otherwise by way of something near happenstance within your particular horizons of privilege, work, and friends and doesn't arrive to it by way of logic or research. In other words, I didn't come to this project through the critique that serves as its introduction in this book; they came after I had committed to lend a hand.

25. In a windowless bar, after I gave a talk that was attended by just one person, the architect Kiel Moe helped me crystallize this language.

26. My role, like that of everyone else, is multiple and ill defined. At times, parsing one's individual contributions to a project like this can feel out of touch and self-interested, because all aspects of the work are brought into being by many hands. At other times, it feels absolutely necessary because volunteer labor can be made invisible in a project of this scale and tacitly credited to the most prominent figures involved. The need for generous attribution is likely inversely correlated with eminence, although much of the world works on an opposite model. In my contributions, I led the solar balloon launch in Antarctica, demonstrating how well the balloons work in remote, resource-scarce regions, and in Idaho during the 2017 solar eclipse to assess the effect of the eclipse on balloon internal temperature and altitude; helped to advance the open-source hardware licensing of the project; and much more.

27. The platform uses public data from the Global Forecast System—maintained by the US National Weather Service—to model the flight trajectories. Just as you can select flexible dates in an airline marketplace to find the cheapest ticket, you can note that your dates are flexible to find out which winds on which days will get you closest to your destination.

28. This attempt to understand this mode of flight as both of the contemporary moment (because it is currently made with high-grade, hydrocarbon-derived synthetic fabrics) and beyond it is in conversation with M Murphy's (2017) work that is "both with and against" contemporary forms of technoscience.

29. For Walter Benjamin, the Parisian arcades crystallize the vaporous fantasies that lie under the surface and proliferate out of large technological projects (Buck-Morss 1991: 254; Benjamin 2002). And indeed, architectural historians note that the invention of cast wrought iron is the "single great technical innovation to take place in building before the twentieth century" (Kronenburg 2002: 45). What is not addressed in the literature is how subsequent construction technologies have maintained or modified the technological enchantment of the built environment in the intervening 160+ years. I would hazard the assertion: If iron in the form of arcades fueled the commercial dreamworlds of the mid-nineteenth century, engineered woods supplied the tinder for the combined commercial and domestic fantasies of the mid-twentieth-century United States. The aerostats in the Grand Palais conjure fantasies that run orthogonal to those of iron and engineered wood.

30. You can donate to the national community wealth network, Seed Commons, at seedcommons.org/get-involved/. You can support national community-owned mobile-home communities at rocusa.org/fund-a-community-project/. You can support Dulce Lomita and other allied projects through their neighborhood nonprofit at poderemma .org/donate.

31. See KC Tenants Union (kctenants.org), Appalachians Against Pipelines (aapsolidarity.org), Firelands (firelandswa.org/), and Rural Organizing Project (rop.org/the -project/).

REFERENCES

Abel, Richard L. 1987. "The Real Tort Crisis—Too Few Claims." *Ohio State Law Journal* 48: 443–67.

Adams, Vincanne. 2013. *Markets of Sorrow, Labors of Faith: New Orleans in the Wake of Katrina*. Durham, NC: Duke University Press.

Adams, Vincanne. 2023. *Glyphosate and the Swirl: An Agroindustrial Chemical on the Move*. Critical Global Health: Evidence, Efficacy, Ethnography. Durham, NC: Duke University Press.

Adams, Vincanne, Taslim van Hattum, and Diana English. 2009. "Chronic Disaster Syndrome: Displacement, Disaster Capitalism, and the Eviction of the Poor from New Orleans." *American Ethnologist* 36, no. 4 (November 1): 615–36. https://doi.org/10.1111/j.1548-1425.2009.01199.x.

Agard-Jones, Vanessa. 2012. "What the Sands Remember." *GLQ: A Journal of Lesbian and Gay Studies* 18, nos. 2–3 (June 1): 325–46. https://doi.org/10.1215/10642684-1472917.

Agard-Jones, Vanessa. 2013. "Bodies in the System." *Small Axe: A Caribbean Journal of Criticism* 17, no. 3 (November 1): 182–92. https://doi.org/10.1215/07990537-2378991.

Ahmed, Sara. 2017. *Living a Feminist Life*. Illustrated ed. Durham, NC: Duke University Press.

Aidala, James. 2010. Chemical Heritage Foundation, Philadelphia, Oral History Transcript #0660. Interview by Jody A. Roberts and Kavita D. Hardy, May 20.

Albright, Thomas D. 2023. "Scientist's Take on Scientific Evidence in the Courtroom." *Proceedings of the National Academy of Sciences* 120, no. 41 (October 10): e2301839120. https://doi.org/10.1073/pnas.2301839120.

Alexander, Diane, and Janet Currie. 2017. "Is It Who You Are or Where You Live? Residential Segregation and Racial Gaps in Childhood Asthma." *Journal of Health Economics* 55 (September 1): 186–200. https://doi.org/10.1016/j.jhealeco.2017.07.003.

Allen, Barbara L. 2003. *Uneasy Alchemy: Citizens and Experts in Louisiana's Chemical Corridor Disputes*. Cambridge, MA: MIT Press.

American Chemistry Council. 2024. "Formaldehyde." American Chemistry Council, May 31. https://www.americanchemistry.com/chemistry-in-america/chemistries/formaldehyde.

Ashford, Nicholas A., and Charles C. Caldart. 1996. *Technology, Law, and the Working Environment*. Rev. ed. Washington, DC: Island Press.

Associated Press. 2006. "Environmental Group Says Some FEMA Trailers Unsafe." May 18.

Associated Press. 2008. "FEMA Offers Refunds for Katrina Trailers." January 17.

ATSDR (Agency for Toxic Substances and Disease Registry). 2007. "Revised Formaldehyde Sampling of FEMA Temporary-Housing Trailers, October, 2007." Atlanta, GA, October. http://www.atsdr.cdc.gov/substances/formaldehyde/public_assessment.html.

ATSDR (Agency for Toxic Substances and Disease Registry). 2014. "Medical Management Guidelines for Formaldehyde." October. http://www.atsdr.cdc.gov/mmg/mmg.asp?id=216&tid=39.

ATSDR (Agency for Toxic Substances and Disease Registry). 2016. "Health Consultation Brooklyn Township PM2.5: Brooklyn Township, Susquehanna County, Pennsylvania Cost Recovery Number: 3A4K00." Atlanta, GA, April 22. https://www.atsdr.cdc.gov/HAC/pha/BrooklynTownship/BrooklynTwnsp_pm2-5_HC_Final_04-22-2016_508.pdf.

Auyero, Javier, and Debora Swistun. 2008. *Flammable: Environmental Suffering in an Argentine Shantytown*. Oxford: Oxford University Press.

Axelrad, Robert. 1992. "EPA Indoor Air Division Robert Axelrad Response Letter." February 24. https://nepis.epa.gov/Exe/ZyPDF.cgi/000002IB.PDF?Dockey=000002IB.pdf.

Bailey, Moya. 2021. *Misogynoir Transformed: Black Women's Digital Resistance*. New York: New York University Press.

Baker, Jim, Liz Voigt, and Linda Jun. 2019. "Private Equity Giants Converge on Manufactured Homes." February. https://pestakeholder.org/wp-content/uploads/2019/02/Private-Equity-Giants-Converge-on-Manufactured-Homes-PESP-MHAction-AFR-021419.pdf.

Baker, Mike, and Daniel Wagner. 2015. "Minorities Exploited by Warren Buffett's Mobile-Home Empire." *Seattle Times*, December 26. https://www.seattletimes.com/seattle-news/times-watchdog/minorities-exploited-by-warren-buffetts-mobile-home-empire-clayton-homes/.

Balayannis, Angeliki, and Emma Garnett. 2020. "Chemical Kinship: Interdisciplinary Experiments with Pollution." *Catalyst: Feminism, Theory, Technoscience* 6, no. 1 (May 15). https://doi.org/10.28968/cftt.v6i1.33524.

Baldwin, James. (1963) 1992. *The Fire Next Time*. Reissue ed. New York: Vintage.

Barber, Daniel A. 2023. *Modern Architecture and Climate: Design Before Air Conditioning*. Princeton, NJ: Princeton University Press.

Barrows, Kari. 2017. "Housing Co-Ops, a Potential Affordable Housing Solution." *Mountain Xpress*, May 8. http://mountainx.com/news/housing-co-ops-a-potential-affordable-housing-solution/.

Baudrillard, Jean. 1995. *Simulacra and Simulation*. Ann Arbor: University of Michigan Press.

Bellacasa, María Puig de la. 2017. *Matters of Care: Speculative Ethics in More Than Human Worlds*. Minneapolis: University of Minnesota Press.

Benjamin, Walter. 2002. *The Arcades Project*. Edited by Rolf Tiedemann. Translated by Howard Eiland and Kevin McLaughlin. Cambridge, MA: Belknap Press of Harvard University Press.

Berlant, Lauren. 2007. "Nearly Utopian, Nearly Normal: Post-Fordist Affect in La Promesse and Rosetta." *Public Culture* 19, no. 2 (March 20): 273–301. https://doi.org /10.1215/08992363-2006-036.

Bernardi, Roy. 2007. "Federal Housing Response to Hurricane Katrina: Written Statement of HUD Deputy Secretary." Washington, DC, February 6.

Berry, Chip, Carolyn Hronis, and Maggie Woodward. 2018. "Who's Energy Insecure? You Might Be Surprised." *ACEEE Proceedings*. https://www.aceee.org/files/proceedings /2018/index.html#/paper/event-data/p393.

Best, Stephen, and Saidiya Hartman. 2005. "Fugitive Justice." *Representations* 92, no. 1: 1–15. https://doi.org/10.1525/rep.2005.92.1.1.

Biehl, João Guilherme. 2007. "Pharmaceuticalization: AIDS Treatment and Global Health Politics." *Anthropological Quarterly* 80, no. 4: 1083–126. https://search .ebscohost.com/login.aspx?direct=true&db=f5h&AN=27881836&site=ehost-live.

Blanchette, Alex. 2019. "Living Waste and the Labor of Toxic Health on American Factory Farms." *Medical Anthropology Quarterly* 33, no. 1: 80–100. https://doi.org/10.1111 /maq.12491.

Bond, David. 2022. *Negative Ecologies: Fossil Fuels and the Discovery of the Environment*. Berkeley: University of California Press.

Boudia, Soraya, Angela N. H. Creager, Scott Frickel, et al. 2021. *Residues: Thinking Through Chemical Environments*. New Brunswick, NJ: Rutgers University Press.

Boudia, Soraya, and Nathalie Jas, eds. 2013. *Toxicants, Health and Regulation Since 1945*. London: Pickering and Chatto.

Boudia, Soraya, and Nathalie Jas. 2014. *Powerless Science? Science and Politics in a Toxic World*. New York: Berghahn.

Bouie, Jamelle. 2015. "If You Want to Understand Black Lives Matter, You Have to Understand Katrina." *Slate*, August 23. http://www.slate.com/articles/news_and_politics /politics/2015/08/hurricane_katrina_10th_anniversary_how_the_black_lives_matter _movement_was.html.

Boullier, Henri, and Emmanuel Henry. 2022. "Toxic Ignorance. How Regulatory Procedures and Industrial Knowledge Jeopardise the Risk Assessment of Chemicals." *Science as Culture* (April 17): 1–24. https://doi.org/10.1080/09505431.2022.2062319.

Boyer, Dominic, and Timothy Morton. 2021. *Hyposubjects: On Becoming Human*. Open Humanities Press. http://www.openhumanitiespress.org/books/titles /hyposubjects/.

Brant, Clare. 2008. "Scenting a Subject: Odour Poetics and the Politics of Space." *Ethnos* 73, no. 4: 544–63. https://doi.org/10.1080/00141840802563964.

Brooks, Paul. 1989. *The House of Life: Rachel Carson at Work*. Boston: Houghton Mifflin.

Brown, Bob, Leon El-Alamin, Latisha Jones, et al. 2020. "A Long Way from Justice: Reflections from Flint on the $600 Million Settlement Proposal." *Environmental Justice* 13, no. 6 (December 1): 222–24. https://doi.org/10.1089/env.2020.0048.

Brown, Phil. 1992. "Popular Epidemiology and Toxic Waste Contamination: Lay and Professional Ways of Knowing." *Journal of Health and Social Behavior* 33, no. 3 (September 1): 267–81. https://doi.org/10.2307/2137356.

Brown, Phil. 1997. "Popular Epidemiology Revisited." *Current Sociology* 45, no. 3 (July 1): 137–56. https://doi.org/10.1177/001139297045003008.

Buck-Morss, Susan. 1991. *The Dialectics of Seeing: Walter Benjamin and the Arcades Project*. Cambridge, MA: MIT Press.

Bullard, Robert D., and Beverly Wright. 2009. *Race, Place, and Environmental Justice After Hurricane Katrina: Struggles to Reclaim, Rebuild, and Revitalize New Orleans and the Gulf Coast*. Boulder, CO: Routledge.

Burkhart, Ann M. 2010. "Bringing Manufactured Housing into the Real Estate Finance System." *Pepperdine Law Review* 37, no. 2: 427–58. https://papers.ssrn.com/sol3/papers.cfm?abstract_id=1548441.

California Air Resources Board. 2007. "Proposed Airborne Toxic Control Measure to Reduce Formaldehyde Emissions from Composite Wood Products." March 9. https://ww3.arb.ca.gov/regact/2007/compwood07/isor.pdf.

Calvillo, Nerea. 2018. "Political Airs: From Monitoring to Attuned Sensing Air Pollution." *Social Studies of Science* 48, no. 3 (June 1): 372–88. https://doi.org/10.1177/0306312718784656.

Calvillo, Nerea. 2023. *Aeropolis: Queering Air in Toxicpolluted Worlds*. New York: Columbia University Press.

Calvillo, Nerea, and Emma Garnett. 2019. "Data Intimacies: Building Infrastructures for Intensified Embodied Encounters with Air Pollution." *Sociological Review* 67, no. 2 (March 1): 340–56. https://doi.org/10.1177/0038026119830575.

Carr, Mariel. 2015. *People Are Still Living in FEMA's Toxic Trailers*. YouTube video, 19:08, September 27. https://www.youtube.com/watch?v=rtj60-cBHQE.

Castle, Emery N. 1995. *The American Countryside: Rural People and Places*. Lawrence: University Press of Kansas.

Cattaneo, Laura, and Donna Feir. 2019. "The Higher Price of Mortgage Financing for Native Americans." Working Paper Series. Center for Indian Country Development, October 3. https://web.archive.org/web/20191118054637/https://www.minneapolisfed.org/~/media/assets/indiancountry/working-papers/cicd-wp-201906.pdf.

CDC (Centers for Disease Control and Prevention). 2010. "Final Report on Formaldehyde Levels in FEMA-Supplied Travel Trailers, Park Models, and Mobile Homes." Atlanta, GA: Centers for Disease Control and Prevention.

Center for Justice and Democracy. 2020. "How Low Can They Go? Civil, Tort, Med Mal, Products Caseloads and Jury Trials." New York Law School, September 15. https://centerjd.org/content/how-low-can-they-go-civil-tort-med-mal-products-caseloads-and-jury-trials.

Chen, Mel Y. 2012. *Animacies: Biopolitics, Racial Mattering, and Queer Affect*. Durham, NC: Duke University Press.

Cheng, Anne Anlin. 2001. *The Melancholy of Race: Psychoanalysis, Assimilation, and Hidden Grief*. Oxford: Oxford University Press.

Chiang, S. Leo. 2010. *A Village Called Versailles*. Documentary film.

Children's Health Fund, National Center for Disaster Preparedness, and Columbia University Mailman School of Public Health. 2008. "Legacy of Shame: The On-Going Public Health Disaster of Children Struggling in Post-Katrina Louisiana." November 4.

Choy, Timothy. 2011. *Ecologies of Comparison: An Ethnography of Endangerment in Hong Kong*. Durham, NC: Duke University Press.

Choy, Timothy K., Lieba Faier, Michael J. Hathaway, Miyako Inoue, Shiho Satsuka, and Anna Tsing. 2009. "A New Form of Collaboration in Cultural Anthropology: Matsutake Worlds." *American Ethnologist* 36, no. 2: 380–403. https://doi.org/10.1111/j.1548-1425.2009.01141.x.

Cifor, M., P. Garcia, T. L. Cowan, et al. 2019. "Feminist Data Manifest-No." https://www.manifestno.com.

Clarke, Kamari Maxine. 2019. *Affective Justice: The International Criminal Court and the Pan-Africanist Pushback*. Durham, NC: Duke University Press.

Clay, Don R. 2012. Science History Institute, Philadelphia, Oral History Transcript # 0684. Interview by Jody A. Roberts and Kavita D. Hardy, March 16.

CLEAR. 2021. *CLEAR Lab Book: A Living Manual of Our Values, Guidelines, and Protocols*. Ver. 3. St. John's: Civic Laboratory for Environmental Action Research, Memorial University of Newfoundland and Labrador. https://civiclaboratory.nl/clear-lab-book/.

Clifford, James. 1981. "On Ethnographic Surrealism." *Comparative Studies in Society and History* 23, no. 4 (October 1): 539–64. https://doi.org/10.2307/178393.

COEHHA (California Office of Environmental Health Hazard Assessment). 2001. "Prioritization of Toxic Air Contaminants—Formaldehyde." Children's Environmental Health Protection Act.

COEHHA (California Office of Environmental Health Hazard Assessment). 2008. "Individual Acute, 8-Hour, and Chronic Reference Exposure Level Summaries. Appendix to 'Air Toxics Hot Spots Program Technical Support Document for the Derivation of Noncancer Reference Exposure Levels.'" December. http://oehha.ca.gov/air/hot_spots/rels_dec2008.html.

Cohen, Ariella. 2010. "Council Wants City to Get Rid of Remaining FEMA Trailers." *The Lens*, September 20. http://thelensnola.org/2010/09/20/fema-triailers/.

Cohen, Thomas H., and Kyle Harbacek. 2011. "Punitive Damage Awards in State Courts, 2005." Washington, DC: US Department of Justice. http://www.bjs.gov/content/pub/pdf/pdasc05.pdf.

Collins, Patricia Hill. 1990. *Black Feminist Thought: Knowledge, Consciousness, and the Politics of Empowerment*. New York: Routledge.

Committee on Homeland Security and Governmental Affairs. 2006. "FEMA's Manufactured Housing Program: Haste Makes Waste." April 21. Field Hearing in Hope, AR. US Government Printing Office. https://bulk.resource.org/gpo.gov/hearings/109s/28240.pdf.

Consumer Reports. 2010. "Natural Air Purifier." https://web.archive.org/web/20110730110529/http://www.consumerreports.org/cro/magazine-archive/2010/september/appliances/best-air-purifier/natural-air-purifier/index.htm.

Cooper, Kenneth J. 2011. "As Housing Shortage Worsens, Tribes Forced to Use FEMA Trailers." *Truthout*, May 6. https://truthout.org/articles/as-housing-shortage-worsens -tribes-forced-to-use-fema-trailers/.

Cordner, Alissa. 2016. *Toxic Safety: Flame Retardants, Chemical Controversies, and Environmental Health*. New York: Columbia University Press.

Covington and Burling. 1995. "'Tort Reform Project Budget' UCSF Truth Tobacco Industry Documents." Philip Morris Records; Master Settlement Agreement. Unknown, October 3. https://www.industrydocuments.ucsf.edu/tobacco/docs/#id=ysgn0085.

Cowie, Elizabeth. 2007. "Ways of Seeing: Documentary Film and the Surreal of Reality." In *Building Bridges: The Cinema of Jean Rouch*, edited by Joram Ten Brink, 201–18. New York: Columbia University Press.

Craemer, Thomas. 2010. "Evaluating Racial Disparities in Hurricane Katrina Relief Using Direct Trailer Counts in New Orleans and FEMA Records." *Public Administration Review* 70, no. 3: 367–77. https://doi.org/10.1111/j.1540-6210.2010.02151.x.

Cram, Shannon. 2016. "Living in Dose: Nuclear Work and the Politics of Permissible Exposure." *Public Culture* 28, no. 3 (September 1): 519–39. https://doi.org/10.1215 /08992363-3511526.

Creative Interventions, ed. 2021. *Creative Interventions Toolkit*. AK Press. https://www .akpress.org/creative-interventions-toolkit.html.

Cree, Delvin. 2012. "FEMA Trailer Problems Still an Issue in Indian Country." *Indianz*. Accessed September 6, 2021. https://www.indianz.com/News/2012/07/06/delvin -cree-fema-trailer-probl.asp.

Critical Resistance. 2021. "Reformist Reforms vs. Abolitionist Steps to End Imprisonment." https://criticalresistance.org/wp-content/uploads/2021/08/CR _abolitioniststeps_antiexpansion_2021_eng.pdf.

Critical Resistance. n.d. "What Is the PIC? What Is Abolition?" Accessed March 20, 2024. https://criticalresistance.org/mission-vision/not-so-common-language/.

CSC (Consumer Safety Commission). 1997. "An Update on Formaldehyde." Washington, DC: Consumer Safety Commission.

Curran, Barbara A. 1977. *The Legal Needs of the Public: The Final Report of a National Survey, a Joint Undertaking by the American Bar Association Special Committee to Survey Legal Needs and the American Bar Foundation*. Chicago: American Bar Foundation.

Cussins, Charis. 1996. "Ontological Choreography: Agency Through Objectification in Infertility Clinics." *Social Studies of Science* 26, no. 3 (August 1): 575–610. https://doi .org/10.1177/030631296026003004.

Dannemiller, Karen C., Johnna S. Murphy, Sherry L. Dixon, et al. 2013. "Formaldehyde Concentrations in Household Air of Asthma Patients Determined Using Colorimetric Detector Tubes." *Indoor Air* 23, no. 4: 285–94. https://doi.org/10.1111/ina.12024.

Danzon, Patricia Munch. 1985. *Medical Malpractice: Theory, Evidence, and Public Policy*. Cambridge, MA: Harvard University Press.

Davies, Thom. 2018. "Toxic Space and Time: Slow Violence, Necropolitics, and Petrochemical Pollution." *Annals of the American Association of Geographers* 108, no. 6 (November 2): 1537–53. https://doi.org/10.1080/24694452.2018.1470924.

Davies, Thom, Arshad Isakjee, and Jelena Obradovic-Wochnik. 2023. "Epistemic Borderwork: Violent Pushbacks, Refugees, and the Politics of Knowledge at the EU Border." *Annals of the American Association of Geographers* 113, no. 1 (January 2): 169–88. https://doi.org/10.1080/24694452.2022.2077167.

Davis, Angela Y. 2003. *Are Prisons Obsolete?* Rev. ed. New York: Seven Stories.

Davis, Matt. 2010. "Five Years Later, a Richer, Whiter New Orleans." *Salon*, August 29. https://www.salon.com/2010/08/28/new_orleans_katrina_demographics/.

Davis, Virginia. 2002. "A Discovery of Sorts: Reexamining the Origins of the Federal Indian Housing Obligation." *Harvard Blackletter Law Journal* 18: 211.

DeLeon, Sara, Rayko Halitschke, Ralph S. Hames, André Kessler, Timothy J. DeVoogd, and André A. Dhondt. 2013. "The Effect of Polychlorinated Biphenyls on the Song of Two Passerine Species." *PLOS ONE* 8, no. 9 (September 18): e73471. https://doi.org/10.1371/journal.pone.0073471.

Dempsey, Jessica. 2013. "Biodiversity Loss as Material Risk: Tracking the Changing Meanings and Materialities of Biodiversity Conservation." *Geoforum* 45 (March 1): 41–51. https://doi.org/10.1016/j.geoforum.2012.04.002.

Dendle, Peter. "The Zombie as a Barometer of Cultural Anxiety." In *Monsters and the Monstrous: Myths and Metaphors of Enduring Evil*, edited by Niall Scott, 45–57. Amsterdam: Rodopi, 2007.

Dillon, Lindsey, Rebecca Lave, Becky Mansfield, et al. 2019. "Situating Data in a Trumpian Era: The Environmental Data and Governance Initiative." *Annals of the American Association of Geographers* 109, no. 2 (March 4): 545–55. https://doi.org/10.1080/24694452.2018.1511410.

Dillon, Lindsey, Christopher Sellers, Vivian Underhill, et al. 2018. "The Environmental Protection Agency in the Early Trump Administration: Prelude to Regulatory Capture." *American Journal of Public Health* 108, no. S2 (April): S89–94. https://doi.org/10.2105/AJPH.2018.304360.

Dixon, Ejeris, and Leah Lakshmi Piepzna-Samarasinha, eds. 2020. *Beyond Survival: Strategies and Stories from the Transformative Justice Movement*. Annotated ed. Chico, CA: AK Press.

Dlamini, Jacob. 2009. *Native Nostalgia*. Jacana Media.

Dominish, E., N. Florin, and S. Teske. 2019. "Responsible Minerals Sourcing for Renewable Energy." Report prepared for Earthworks by the Institute for Sustainable Futures, University of Technology Sydney. https://earthworks.org/publications/responsible-minerals-sourcing-for-renewable-energy/.

Donohue, Allie, Nicholas Shapiro, Elizabeth Lara, and Karen Dannemiller. 2019. "Assessment of the Microbiome of Dust Before and After Hydroponics System Installation in Homes." Poster, Ohio State University, College of Engineering.

Downey, Gary Lee, and Teun Zuiderent-Jerak, eds. 2021. *Making and Doing: Activating STS Through Knowledge Expression and Travel*. Cambridge, MA: MIT Press.

du Gouw, J. A., S. A. McKeen, K. C. Aikin, et al. 2015. "Airborne Measurements of the Atmospheric Emissions from a Fuel Ethanol Refinery." *Journal of Geophysical Research: Atmospheres* 120, no. 9: 4385–97. https://doi.org/10.1002/2015JD023138.

Earthseed. n.d. "Our Collective." Earthseed Land Collective. Accessed November 16, 2024. https://earthseedlandcoop.org/about/.

Edmunds, David S., Ryan Shelby, Angela James, et al. 2013. "Tribal Housing, Codesign, and Cultural Sovereignty." *Science, Technology, and Human Values* 38, no. 6 (November 1): 801–28. https://doi.org/10.1177/0162243913490812.

Einstein, Carl. 1929. "Andre Masson, Etude Ethnologique." *Documents* 1, no. 2: 95.

Ellis, Erle C. 2023. "Why I Resigned from the Anthropocene Working Group—Anthroecology Lab." July 13. https://anthroecology.org/why-i-resigned-from-the-anthropocene-working-group/.

Engber, Daniel. 2006. "Who Unlocked My Trailer? FEMA Gave Us All the Same Key." *Slate*, August 15.

Engel, David M. 1984. "The Oven Bird's Song: Insiders, Outsiders, and Personal Injuries in an American Community." *Law and Society Review* 18, no. 4: 551. https://doi.org/10.2307/3053447.

Engelmann, Sasha. 2020. *Sensing Art in the Atmosphere: Elemental Lures and Aerosolar Practices*. London: Routledge.

Engelmann, Sasha, and Derek McCormack. 2018. "Elemental Aesthetics: On Artistic Experiments with Solar Energy." *Annals of the American Association of Geographers* 108, no. 1 (January 2): 241–59. https://doi.org/10.1080/24694452.2017.1353901.

Engelmann, Sasha, and Derek McCormack. 2021. "Elemental Worlds: Specificities, Exposures, Alchemies." *Progress in Human Geography* 45, no. 6 (December 1): 1419–39. https://doi.org/10.1177/0309132520987301.

EPA (US Environmental Protection Agency). 1975. "Quarles Testifies on the Need for Toxic Substances Act." Speeches, Testimony and Transcripts, July 10. http://www2.epa.gov/aboutepa/quarles-testifies-need-toxic-substances-act.

EPA (US Environmental Protection Agency). 1987. "The Total Exposure Assessment Methodology (TEAM) Study: Summary and Analysis." Washington, DC.

EPA (US Environmental Protection Agency). 2024. "IRIS Toxicological Review of Formaldehyde (Inhalation)." Washington, DC.

EPA (US Environmental Protection Agency). 2025. "2020 AirToxScreen: Assessment Results." January 2. https://www.epa.gov/AirToxScreen/2020-airtoxscreen-assessment-results.

Estes, Nick. 2019. *Our History Is the Future: Standing Rock Versus the Dakota Access Pipeline, and the Long Tradition of Indigenous Resistance*. New York: Verso.

Estes, Nick, Ruth Wilson Gilmore, and Christopher Loperena. 2021. "United in Struggle: As Racial Capitalism Rages, Movements for Indigenous Sovereignty and Abolition Offer Visions of Freedom on Stolen Land." *NACLA Report on the Americas* 53, no. 3 (July 3): 255–67. https://doi.org/10.1080/10714839.2021.1961444.

Fassin, Didier, and Mariella Pandolfi. 2010. *Contemporary States of Emergency: The Politics of Military and Humanitarian Interventions*. New York: Zone Books.

FEMA (Federal Emergency Management Agency). 2007. "Transfer of FEMA Unused Mobile Homes to Tribal Governments." http://www.fema.gov/media/archives/2007/091207.shtm.

Feser, Alison. 2020. "Reproducing Photochemical Life in the Imaging Capital of the World." PhD diss., University of Chicago. https://www.proquest.com/docview /2427529922/abstract/532BF6BA15FE4745PQ/1.

Flaherty, Jordan. 2010. *Floodlines: Community and Resistance from Katrina to the Jena Six.* Chicago: Haymarket Books.

Fletcher, Christopher. 2005. "Dystoposthesia: Emplacing Environmental Sensitivities." In *Empire of the Senses: The Sensual Culture Reader,* edited by David Howes, 380–98. Oxford: Berg.

Fortun, Kim. 2001. *Advocacy After Bhopal: Environmentalism, Disaster, New Global Orders.* Chicago: University of Chicago Press.

Fortun, Kim. 2003. "Ethnography in/of/as Open Systems." *Reviews in Anthropology* 32, no. 2: 171–90.

Fortun, Kim. 2012. "Ethnography in Late Industrialism." *Cultural Anthropology* 27, no. 3 (August): 446–64. https://doi.org/10.1111/j.1548-1360.2012.01153.x.

Fortun, Kim, and Mike Fortun. 2005. "Scientific Imaginaries and Ethical Plateaus in Contemporary U.S. Toxicology." *American Anthropologist* 107, no. 1: 43–54. https:// doi.org/10.1525/aa.2005.107.1.043.

Foucault, Michel. 2003. *"Society Must Be Defended": Lectures at the Collège de France, 1975–1976.* New York: Picador.

Freudenberg, Nicholas. 1984. *Not in Our Backyards! Community Action for Health and the Environment.* New York: New York University Press.

Frickel, Scott. 2008. "On Missing New Orleans: Lost Knowledge and Knowledge Gaps in an Urban Hazardscape." *Environmental History* 13, no. 4: 643–50.

Frihart, Charles R., Timothy L. Chaffee, and James M. Wescott. 2020. "Long-Term Formaldehyde Emission Potential from UF- and NAF-Bonded Particleboards." *Polymers* 12, no. 8 (August): 1852. https://doi.org/10.3390/polym12081852.

GAO (US Government Accountability Office). 1991. "[Title Obscured] B-244536." August 1.

GAO (US Government Accountability Office). 2008. "Chemical Assessments: EPA's New Assessment Process Will Further Limit the Productivity and Credibility of Its Integrated Risk Information System." May 21. https://www.gao.gov/assets/gao-08-810t.pdf.

Garrett, M. H., M. A. Hooper, B. M. Hooper, P. R. Rayment, and M. J. Abramson. 1999. "Increased Risk of Allergy in Children Due to Formaldehyde Exposure in Homes." *Allergy* 54, no. 4 (April): 330–37.

Garry, Vincent. 2008. "Formaldehyde & FEMA Provided Mobile Home Trailers: Letter to I&O Subcommittee Chairman Brad Miller." March 24.

Gartman, David. 2004. "Tough Guys and Pretty Boys: The Triumph of Stylists in the 1950s." *Automobile in American Life and Society.* http://autolife.umd.umich.edu /Design/Gartman/D_Casestudy/D_Casestudy7.htm.

Gehl, Lynn, and Algonquin Anishinaabe-kwe. 2012. "Ally Bill of Responsibilities." https://www.lynngehl.com/uploads/5/0/0/4/5004954/ally_bill_of_responsibilities _poster.pdf.

Gehrke, Gretchen, and Nicholas Shapiro. 2015. "Formaldehyde Measurement: Testing Public Lab's Kit with DOH's Equipment." Public Lab, October 7. https://publiclab.org /n/12278.

Gertz, Emily, and Patrick Di Justo. 2012. *Environmental Monitoring with Arduino: Building Simple Devices to Collect Data About the World Around Us*. Sebastopol, CA: Make Community LLC.

Gilio-Whitaker, Dina. 2022. "Environmental Justice Is Only the Beginning." *High Country News*, July 1. https://www.hcn.org/issues/54.7/indigenous-affairs-perspective-environ mental-justice-is-only-the-beginning.

Gilmore, Ruth Wilson. 2002. "Race and Globalization." In *Geographies of Global Change: Remapping the World*, edited by Ron J. Johnston, Peter J. Taylor, and Michael J. Watts, 2nd ed. New York: Wiley-Blackwell.

Girman, John R. 1992. "Comment on the Use of Plants as a Means to Control Indoor Air Pollution." Washington, DC: Environmental Protection Agency. https://nepis.epa .gov/Exe/ZyPDF.cgi/000002IB.PDF?Dockey=000002IB.PDF.

Good, Byron J., and Mary-Jo DelVecchio Good. 1994. "In the Subjunctive Mode: Epilepsy Narratives in Turkey." *Social Science and Medicine* 38, no. 6: 835–42.

Goodman, Amy. 2006. "FEMA's Dirty Little Secret: A Rare Look Inside the Renaissance Village Trailer Park, Home to Over 2,000 Hurricane Katrina Evacuees." *Democracy Now!*, April 24. http://www.democracynow.org/2006/4/24/femas_dirty_little_secret_a_rare.

Gorz, André. 1967. *Strategy for Labor: A Radical Proposal*. Boston: Beacon.

Gould, Kenneth A., David N. Pellow, and Allan Schnaiberg. 2015. *Treadmill of Production: Injustice and Unsustainability in the Global Economy*. New York: Routledge.

Grabbs, J., R. Corsi, and V. Torres. 2000. "Volatile Organic Compounds in New Automobiles: Screening Assessment." *Journal of Environmental Engineering* 126, no. 10: 974–77. https://doi.org/10.1061/(ASCE)0733–9372(2000)126:10(974).

Gramsci, Antonio. (1947) 1991. *Prison Notebooks*. Vol. 1. Edited by Joseph A. Buttigieg. New York: Columbia University Press.

Grandia, Liza. 2020. "Carpet Bombings: A Drama of Chemical Injury in Three Acts." *Catalyst: Feminism, Theory, Technoscience* 6, no. 1. https://catalystjournal.org/index .php/catalyst/article/view/32532.

Grealy, Liam. 2023. "Governing Disassembly in Indigenous Housing." *Housing Studies* 38, no. 2: 327–46. https://doi.org/10.1080/02673037.2021.1882662.

Grealy, Liam, Tess Lea, Megan Moskos, Richard Benedict, Daphne Habibis, and Stephanie King. 2022. "Sustaining Housing Through Planned Maintenance in Remote Central Australia." *Housing Studies*, 1–23. https://doi.org/10.1080/02673037.2022 .2084045.

Green, Reuben Keith. 2020. "The Case for Renaming the USS John C. Stennis." *US Naval Institute*, June 22. https://www.usni.org/magazines/proceedings/2020/june /case-renaming-uss-john-c-stennis.

Guthman, Julie. 2011. *Weighing In: Obesity, Food Justice, and the Limits of Capitalism*. Berkeley: University of California Press.

Hage, Ghassan. 2015. *Alter-Politics: Critical Anthropology and the Radical Imagination*. Carlton, Victoria: Melbourne University Publishing.

Hale, Charles. 2006. "Activist Research v. Cultural Critique: Indigenous Land Rights and the Contradictions of Politically Engaged Anthropology." *Cultural Anthropology* 21, no 1: 96–120.

Hameed, Irfan, Socrates Kakoulides, John Moscona, et al. 2011. "Acute Myocardial Infarction Before and After the Storm: Hurricane Katrina." *Journal of the American College of Cardiology* 57, no. 14S1 (April 5): E2018. https://doi.org/10.1016/S0735 -1097(11)62018-4.

Hamraie, Aimi. 2017. *Building Access: Universal Design and the Politics of Disability.* 3rd ed. Minneapolis: University of Minnesota Press.

Haraway, Donna Jeanne. 2016. *Staying with the Trouble: Making Kin in the Chthulucene.* Experimental Futures: Technological Lives, Scientific Arts, Anthropological Voices. Durham, NC: Duke University Press.

Harding, Sandra G. 2004. *The Feminist Standpoint Theory Reader: Intellectual and Political Controversies.* New York: Routledge.

Harney, Stefano, and Fred Moten. 2013. *The Undercommons: Fugitive Planning and Black Study.* Brooklyn, NY: Autonomedia.

Hartman, Saidiya. 1997. *Scenes of Subjection: Terror, Slavery, and Self-Making in Nineteenth-Century America.* New York: Oxford University Press.

Henderson, E., and B. Gaston. 2005. "SNOR and Wheeze: The Asthma Enzyme?" *Trends in Molecular Medicine* 11, no. 11: 481–84. https://doi.org/10.1016/j.molmed .2005.09.009.

Hepler-Smith, Evan. 2019. "Molecular Bureaucracy: Toxicological Information and Environmental Protection." *Environmental History* 24, no. 3: 534–60. https://doi.org /10.1093/envhis/emy134.

Hermer, Joe, and Alan Hunt. 1996. "Official Graffiti of the Everyday." *Law and Society Review* 30, no. 3: 455–80. https://doi.org/10.2307/3054125.

Hernandez, R. R., S. B. Easter, M. L. Murphy-Mariscal, et al. 2014. "Environmental Impacts of Utility-Scale Solar Energy." *Renewable and Sustainable Energy Reviews* 29: 766–79. https://doi.org/10.1016/j.rser.2013.08.041.

Herrmann, Anne, and Abigail Stewarts. 2001. "The Combahee River Collective Statement." In *Theorizing Feminism*, 2nd ed., 29–37. New York: Routledge.

Holton, Graham. 2022. "Indigenous Rights Take Center Stage in Chile's New Constitution." *People's World*, July 20. https://peoplesworld.org/article/indigenous-rights-take -center-stage-in-chiles-new-constitution/.

Hot Coffee. 2011. Documentary film. Hot Coffee, If Not Now Productions, The Group Entertainment.

Howell, Junia, and James R. Elliott. 2018. "As Disaster Costs Rise, So Does Inequality." *Socius* 4. https://doi.org/10.1177/2378023118816795.

Huber, Matthew T. 2013. *Lifeblood: Oil, Freedom, and the Forces of Capital.* Minneapolis: University of Minnesota Press.

Huson, Freda, and Anne Spice. 2019. "Heal the People, Heal the Land: An Interview with Freda Huson." In *Standing with Standing Rock: Voices from the #NoDAPL Movement*, edited by Nick Estes and Jaskiran Dhillon, 211–21. Minneapolis: University of Minnesota Press.

Indigenous Environmental Network (IEN). 2018. "Indigenous Principles of Just Transition: Responsibility and Relationship, Sovereignty, Transformation for Action." https:// www.ienearth.org/wp-content/uploads/2018/08/PrinciplesJustTransition-BW.pdf.

IPBES (Intergovernmental Science-Policy Platform on Biodiversity and Ecosystem Services). 2019. "Global Assessment Report on Biodiversity and Ecosystem Services of the Intergovernmental Science-Policy Platform on Biodiversity and Ecosystem Services." May 4. https://doi.org/10.5281/ZENODO.3831673.

Isenberg, Nancy. 2016. *White Trash: The 400-Year Untold History of Class in America.* New York: Viking.

Jackson, Deborah Davis. 2011. "Scents of Place: The Dysplacement of a First Nations Community in Canada." *American Anthropologist* 113, no. 4: 606–18.

Jackson, Debra, and Judy Mannix. 2004. "Giving Voice to the Burden of Blame: A Feminist Study of Mothers' Experiences of Mother Blaming." *International Journal of Nursing Practice* 10, no. 4: 150–58. https://doi.org/10.1111/j.1440-172X.2004.00474.x.

Jackson, John Brinckerhoff. *A Sense of Place, a Sense of Time.* New Haven, CT: Yale University Press, 1994.

Jacobsen, Grant D., and Matthew J. Kotchen. 2010. "Are Building Codes Effective at Saving Energy? Evidence from Residential Billing Data in Florida." Working Paper Series. National Bureau of Economic Research, July. https://doi.org/10.3386/w16194.

Jain, Lochlann. 2006. *Injury: The Politics of Product Design and Safety Law in the United States.* Princeton, NJ: Princeton University Press.

Jaworski, Sophia. 2022. "Chemical Disability and Technoscientific Experimental Subjecthood: Reimagining the Canary in the Coal Mine Metaphor." *Catalyst: Feminism, Theory, Technoscience* 8, no. 2 (November 7). https://doi.org/10.28968/cftt.v8i2.36206.

J.D. Power. 2017. "J.D. Power 2017 China Initial Quality Study." September 26. https://www.jdpower.com/business/press-releases/jd-power-2017-china-initial-quality-study.

Jenkins, Destin, and Justin Leroy, eds. 2021. *Histories of Racial Capitalism.* New York: Columbia University Press.

Jing Lu, Junye Miao, Tao Su, Ying Liu, and Rongqiao He. 2013. "Formaldehyde Induces Hyperphosphorylation and Polymerization of Tau Protein Both *in Vitro* and *in Vivo.*" *Biochimica et Biophysica Acta (BBA)—General Subjects* 1830, no. 8: 4102–16. https://doi.org/10.1016/j.bbagen.2013.04.028.

Johnson, Kevin, and Richard M. Todd. 2017. "The Market for Manufactured Home Loans to American Indian and Alaska Native Borrowers in Indian Country Remains Highly Concentrated." Center for Indian Country Development, December 1. https://www.minneapolisfed.org/indiancountry/research-and-articles/cicd-blog/the-market-for-manufactured-home-loans-to-american-indian-and-alaska-native-borrowers-in-indian-country-remains-highly-concentrated.

Kaba, Mariame. 2014. "Police 'Reforms' You Should Always Oppose." *Truthout*, December 7. https://truthout.org/articles/police-reforms-you-should-always-oppose/.

Kaba, Mariame. 2018. "Questions I Regularly Ask Myself When I'm Outraged About Injustice." Twitter, May 26. https://twitter.com/prisonculture/status/1000413472356696065?ref_src=twsrc%5Etfw.

Kaba, Mariame. 2021. *We Do This 'Til We Free Us: Abolitionist Organizing and Transforming Justice.* Chicago: Haymarket Books.

Kaba, Mariame, and Shira Hassan. 2019. *Fumbling Towards Repair: A Workbook for Community Accountability Facilitators*. Workbook ed. Project NIA.

Kaden, Debra A., Corinne Mandin, Gunnar D. Nielsen, and Peder Wolkoff. 2010. "Formaldehyde." In *WHO Guidelines for Indoor Air Quality: Selected Pollutants*. Geneva: World Health Organization. https://www.ncbi.nlm.nih.gov/books/NBK138711/.

Kear, Mark, Margaret Wilder, Patricia Solís, David Hondula, and Mark Bernstein. 2020. "Self-Isolating from COVID-19 in a Mobile Home? That Could Be Deadly in Arizona." *Arizona Republic*, May 3. https://www.azcentral.com/story/opinion/op-ed/2020/05/03/arizona-heat-could-kill-self-isolating-coronavirus-mobile-homes/3043693001/.

Keirns, Carla C. 2004. "Short of Breath: A Social and Intellectual History of Asthma in the United States." PhD diss., University of Pennsylvania.

Kelley, Robin D. G. 2002. *Freedom Dreams: The Black Radical Imagination*. Boston: Beacon.

Kenner, Alison. 2018. *Breathtaking: Asthma Care in a Time of Climate Change*. Minneapolis: University of Minnesota Press.

Kenner, Alison. 2021. "Emplaced Care and Atmospheric Politics in Unbreathable Worlds." *Environment and Planning C: Politics and Space* 39, no. 6: 1113–28. https://doi.org/10.1177/2399654419851347.

Khoder, M. I., A. A. Shakour, S. A. Farag, and A. A. Abdel Hameed. 2000. "Indoor and Outdoor Formaldehyde Concentrations in Homes in Residential Areas in Greater Cairo." *Journal of Environmental Monitoring* 2, no. 2: 123–26. https://doi.org/10.1039/A908756G.

Kincaid, Linda, and Bruce Offermann. 2010. "Unintended Consequences: Formaldehyde Exposures in Green Homes." *The Synergist*, February, 31–34.

Klein, Naomi. 2007. *The Shock Doctrine: The Rise of Disaster Capitalism*. London: Penguin.

Konsmo, Erin Marie, and Karyn Recollet. 2019. "Afterword: Meeting the Land(s) Where They Are At." In *Indigenous and Decolonizing Studies in Education: Mapping the Long View*, edited by Linda Tuhiwai Smith, Eve Tuck, and K. Wayne Yang. New York: Routledge.

Kronenburg, Robert. 2002. *Houses in Motion*. London: Academy Editions.

Krug, Aubrey Streit, Timothy E. Crews, and Thomas P. McKenna. 2022. "A Social Perennial Vision for the North American Great Plains Rooted in the Resilience of a Natural System-Inspired Agriculture." In *Creating Resilient Landscapes in an Era of Climate Change*. New York: Routledge.

Kusenbach, Margarethe. 2009. "Salvaging Decency: Mobile Home Residents' Strategies of Managing the Stigma of 'Trailer' Living." *Qualitative Sociology* 32, no. 4: 399–428. https://doi.org/10.1007/s11133-009-9139-z.

LaDuke, Winona. 1999. *All Our Relations: Native Struggles for Land and Life*. Cambridge, MA: South End Press.

Lancet. 2006. "A Plea to Abandon Asthma as a Disease Concept." *Lancet* 368, no. 9537 (September): P705 https://doi.org/10.1016/S0140-6736(06)69257-X.

Land Based Jawns. n.d. "About—Land Based Jawns." Accessed November 16, 2024. https://landbasedjawns.com/about/.

Landecker, Hannah. 2011. "Food as Exposure: Nutritional Epigenetics and the New Metabolism." *BioSocieties* 6, no. 2: 167–94. https://doi.org/10.1057/biosoc.2011.1.

Landecker, Hannah. 2019. "From Cheap Chicken to Genome Instability: Arsenic, Diabetes, and the Social Nature of One-Carbon Metabolism." *American Journal of Physical Anthropology*, 168: 135. https://hero.epa.gov/hero/index.cfm/reference/details/reference_id/7454051.

Larrance, Ryan, Michael Anastario, and Lynn Lawry. 2007. "Health Status Among Internally Displaced Persons in Louisiana and Mississippi Travel Trailer Parks." *Annals of Emergency Medicine* 49, no. 5: 590–601. https://doi.org/10.1016/j.annemergmed.2006.12.004.

Lea, Tess. 2008. *Bureaucrats and Bleeding Hearts: Indigenous Health in Northern Australia*. Sydney: UNSW Press.

Lea, Tess. 2012. "When Looking for Anarchy, Look to the State: Fantasies of Regulation in Forcing Disorder Within the Australian Indigenous Estate." *Critique of Anthropology* 32, no. 2: 109–24. https://doi.org/10.1177/0308275X12438251.

Lea, Tess, and Paul Pholeros. 2010. "This Is Not a Pipe: The Treacheries of Indigenous Housing." *Public Culture* 22, no. 1: 187–209. https://doi.org/10.1215/08992363-2009-021.

Lee, Dalbyul, and Juchul Jung. 2014. "The Growth of Low-Income Population in Floodplains: A Case Study of Austin, TX." *KSCE Journal of Civil Engineering* 18, no. 2: 683–93. https://doi.org/10.1007/s12205-014-0205-z.

Lefève, Céline, François Thoreau, and Alexis Zimmer. 2020. *Les humanités médicales. L'engagement des sciences humaines et sociales en médecine*. Arcueil: Doin Editeurs.

Leng, Jiapeng, Chih-Wei Liu, Hadley J. Hartwell, et al. 2019. "Evaluation of Inhaled Low-Dose Formaldehyde-Induced DNA Adducts and DNA–Protein Cross-Links by Liquid Chromatography–Tandem Mass Spectrometry." *Archives of Toxicology* 93, no. 3: 763–73. https://doi.org/10.1007/s00204-019-02393-x.

Lerner, Sharon. 2021. "Formaldehyde Causes Leukemia, According to EPA Assessment Suppressed by Trump Officials." *The Intercept*. https://theintercept.com/2021/08/19/formaldehyde-leukemia-epa-trump-suppressed/.

Lerner, Sharon, and Al Shaw. 2024. "Formaldehyde Causes More Cancer Than Any Other Toxic Air Pollutant. Little Is Being Done to Curb the Risk." *ProPublica*. https://www.propublica.org/article/formaldehyde-epa-trump-public-health-danger.

Lezaun, Javier, and Steve Woolgar. 2013. "The Wrong Bin Bag: A Turn to Ontology in Science and Technology Studies?" *Social Studies of Science* 43, no. 4: 321–40. https://www.jstor.org/stable/48646311.

Liboiron, Max. 2016. "Redefining Pollution and Action: The Matter of Plastics." *Journal of Material Culture* 21, no. 1: 87–110. https://doi.org/10.1177/1359183515622966.

Liboiron, Dr Max [@MaxLiboiron]. 2019a. "#LRF2019 Keynote by @evetuck on Research on Our Terms. Featuring Many Pictures of Seals. And Wisdom That Burns. A Thread from a Listener: Https://T.Co/wKkB5et9t1." Twitter, May 2. https://twitter.com/MaxLiboiron/status/1123757798125395971.

Liboiron, Max. 2019b. "'Research on Our Own Terms' by Eve Tuck: A Thread by a Listener." *CLEAR* (blog), May 3. https://civiclaboratory.nl/2019/05/03/research-on-our-own-terms-by-eve-tuck-a-thread-by-a-listener/.

Liboiron, Max. 2021. *Pollution Is Colonialism*. Durham, NC: Duke University Press.

Liboiron, Max, Justine Ammendolia, Katharine Winsor, et al. 2017. "Equity in Author Order: A Feminist Laboratory's Approach." *Catalyst* 3, no. 2. https://doi.org/10.28968/cftt.v3i2.204.

Liboiron, Max, Emily Simmonds, Edward Allen, et al. 2021. "Doing Ethics with Cod." In *Making and Doing: Activating STS Through Knowledge Expression and Travel*. Cambridge, MA: MIT Press.

Liboiron, Max, Manuel Tironi, and Nerea Calvillo. 2018. "Toxic Politics: Acting in a Permanently Polluted World." *Social Studies of Science* 48, no. 3: 331–49. https://doi.org/10.1177/0306312718783087.

Lindström, Martin. 2005. *Brand Sense: How to Build Powerful Brands Through Touch, Taste, Smell, Sight and Sound*. London: Kogan Page.

Lippencott, Mathew. 2015. "Borrowing a Formaldehyde Kit, Take 1." Public Lab, June 25. https://publiclab.org/n/12003.

Litman, Todd. 2006. "Lessons from Katrina and Rita: What Major Disasters Can Teach Transportation Planners." *Journal of Transportation Engineering* 132, no. 1: 11–18. https://doi.org/10.1061/(ASCE)0733-947X(2006)132:1(11).

Luxemburg, Rosa. 2008. *The Essential Rosa Luxemburg: Reform or Revolution and the Mass Strike*. Edited by Helen Scott. Chicago: Haymarket Books.

Lynch, Michael, Simon A. Cole, Ruth McNally, and Kathleen Jordan. 2008. *Truth Machine: The Contentious History of DNA Fingerprinting*. Chicago: University of Chicago Press.

Lyotard, Jean-François. (1983) 1988. *Le Différend: Phrases in Dispute*. Translated by Georges Van Den Abbeele. Minneapolis: University of Minnesota Press.

Macdonald, Isabel. 2011. "The Shelters That Clinton Built." *The Nation*, July 11. http://www.thenation.com/article/161908/shelters-clinton-built.

MacDonald, John D. 1955. *Cry Hard Cry Fast*. New York: Gold Medal.

Macey, Gregg P., Ruth Breech, Mark Chernaik, et al. 2014. "Air Concentrations of Volatile Compounds Near Oil and Gas Production: A Community-Based Exploratory Study." *Environmental Health* 13, no. 1: 82. https://doi.org/10.1186/1476-069X-13-82.

Maddalena, Randy, Marion Russell, Douglas P. Sullivan, and Michael G. Apte. 2008. "Aldehyde and Other Volatile Organic Chemical Emissions in Four FEMA Temporary Housing Units? Final Report." Ernest Orlando Lawrence Berkeley National Laboratory, Berkeley, CA.

Maddalena, Randy, Marion Russell, Douglas P. Sullivan, and Michael G. Apte. 2009. "Formaldehyde and Other Volatile Organic Chemical Emissions in Four FEMA Temporary Housing Units." *Environmental Science and Technology* 43, no. 15: 5626–32. https://doi.org/10.1021/es9011178.

Madlingozi, Tshepo. 2007. "Post-Apartheid Social Movements and the Quest for the Elusive 'New' South Africa." *Journal of Law and Society* 34: 77–98. https://doi.org/10.1111/j.1467-6478.2007.00383.x.

Mah, Alice. 2023. *Petrochemical Planet: Multiscalar Battles of Industrial Transformation*. Durham, NC: Duke University Press.

Majority Staff Analysis. 2008. "Trailer Manufacturers and Elevated Formaldehyde Levels." United States House of Representatives, Committee on Oversight and Government Reform, July 9.

Marx, Karl. (1867) 1990. *Capital*. Vol. 1, *A Critique of Political Economy*. Translated by Ben Fowkes. New York: Penguin Classics.

Masco, Joseph. 2004. "Nuclear Technoaesthetics: Sensory Politics from Trinity to the Virtual Bomb in Los Alamos." *American Ethnologist* 31, no. 3: 349–73. https://doi.org/10.1525/ae.2004.31.3.349.

Masco, Joseph. 2017. "The Crisis in Crisis." *Current Anthropology* 58, no. S15 (November 9): S65–76. https://doi.org/10.1086/688695.

Maynard, Micheline. 2009. "Is Happiness Still That New Car Smell?" *New York Times*, October 22. http://www.nytimes.com/2009/10/22/automobiles/autospecial2/22CHANGE.html.

Mazo, Adam, and Ben Pender-Cudlip. 2018. *Dawnland*. Documentary. Upstander Project, Principle Pictures, Unrendered Films.

McGoey, L. 2007. "On the Will to Ignorance in Bureaucracy." *Economy and Society* 36, no. 2: 212–35. https://doi.org/10.1080/03085140701254282.

Meikle, Jeffrey L. 1995. *American Plastic: A Cultural History*. New Brunswick, NJ: Rutgers University Press.

Melamed, Jodi. 2015. "Racial Capitalism." *Critical Ethnic Studies* 1, no. 1: 76–85. https://doi.org/10.5749/jcritethnstud.1.1.0076.

Merry, Sally Engle. 1990. *Getting Justice and Getting Even: Legal Consciousness Among Working-Class Americans*. Chicago: University of Chicago Press.

Metzl, Jonathan M. 2020. *Dying of Whiteness: How the Politics of Racial Resentment Is Killing America's Heartland*. Updated ed. New York: Basic Books.

Michaels, David. 2008. *Doubt Is Their Product: How Industry's Assault on Science Threatens Your Health*. New York: Oxford University Press.

Miller, DeMond Shondell, and Jason David Rivera. 2010. "Landscapes of Disaster and Place Orientation in the Aftermath of Hurricane Katrina." In *The Sociology of Katrina: Perspectives on a Modern Catastrophe*, edited by David L. Brunsma, David Overfelt, and Steven J. Picou, 177–90. Lanham, MD: Rowman and Littlefield.

Miller, J. David, and David R. McMullin. 2014. "Fungal Secondary Metabolites as Harmful Indoor Air Contaminants: 10 Years On." *Applied Microbiology and Biotechnology* 98, no. 24 (December 1): 9953–66. https://doi.org/10.1007/s00253-014-6178-5.

Minority Staff Report. 2008. "Toxic Trailers—Toxic Lethargy: How the Centers for Disease Control and Prevention Has Failed to Protect Human Health." Committee on Science and Technology, Subcommittee on Investigations and Oversight. US House of Representatives, September.

Mitman, Gregg. 2007. *Breathing Space: How Allergies Shape Our Lives and Landscapes*. New Haven, CT: Yale University Press.

Moe, Kiel. 2015. "Saraceno's Model of Models: The Magnificence of Aerocene." *Aerocene*. http://aerocene.org/newspaper/moe/.

Moran-Thomas, Amy. 2017. "Glucometer Foils." *Limn* 9 (November): 74–81. https://limn.it/articles/glucometer-foils/.

Moore, Jason W. 2015. *Capitalism in the Web of Life: Ecology and the Accumulation of Capital*. New York: Verso.

Moore, Lorrie. 2010. *A Gate at the Stairs*. Reprint. New York: Vintage.

Moscona, John, Sumit Tiwari, Kartik Munshi, Sudesh Srivastav, Patrice Delafontaine, and Anand Irimpen. 2012. "The Effects of Hurricane Katrina on Acute Myocardial Infarction Five Years After the Storm." *Journal of the American College of Cardiology* 59, no. 13, Supplement (March 27): E354. https://doi.org/10.1016/S0735-1097(12)60355-6.

Moss, David. 1999. "Courting Disaster? The Transformation of Federal Disaster Policy Since 1803." In *The Financing of Catastrophe Risk*, 307–62. Chicago: University of Chicago Press.

Müller, Birgit. 2021. "Glyphosate—a Love Story. Ordinary Thoughtlessness and Response-Ability in Industrial Farming." *Journal of Agrarian Change* 21, no. 1: 160–79. https://doi.org/10.1111/joac.12374.

Mulvaney, Dustin. 2014. "Solar Energy Isn't Always as Green as You Think." *IEEE Spectrum: Technology, Engineering, and Science News*. November 13. https://spectrum.ieee.org/green-tech/solar/solar-energy-isnt-always-as-green-as-you-think.

Mundt, Kenneth A., Alexa E. Gallagher, Linda D. Dell, Ethan A. Natelson, Paolo Boffetta, and P. Robinan Gentry. 2017. "Does Occupational Exposure to Formaldehyde Cause Hematotoxicity and Leukemia-Specific Chromosome Changes in Cultured Myeloid Progenitor Cells?" *Critical Reviews in Toxicology* 47, no. 7 (August 9): 598–608. https://doi.org/10.1080/10408444.2017.1301878.

Murphy, Michelle. 2004. "Uncertain Exposures and the Privilege of Imperception: Activist Scientists and Race at the U.S. Environmental Protection Agency." *Osiris* 19: 266–82. http://www.jstor.org/stable/3655244.

Murphy, Michelle. 2006. *Sick Building Syndrome and the Problem of Uncertainty: Environmental Politics, Technoscience, and Women Workers*. Durham, NC: Duke University Press.

Murphy, Michelle. 2007. "Allergy: The History of a Modern Malady." *Social History of Medicine* 20, no. 1 (April 1): 188–89. https://doi.org/10.1093/shm/hkm024.

Murphy, Michelle. 2008a. "Chemical Regimes of Living." *Environmental History* 13, no. 4: 695–703. http://www.jstor.org/stable/25473297.

Murphy, Michelle. 2008b. "FORUM Chemical Regimes of Living." *Environmental History* 13, no. 4. http://www.historycooperative.org/cgi-bin/justtop.cgi?act=justtop&url=http://www.historycooperative.org/journals/eh/13.4/murphy.html.

Murphy, Michelle. 2010. "Scale, Topography, Origami." Presented at Scalography—An International Workshop, University of Oxford, July 8.

Murphy, Michelle. 2013. "Chemical Infrastructures of the St. Clair River." In *Toxicants, Health and Regulation Since 1945*, edited by Nathalie Jas and Soraya Boudia, 103–15. New York: Routledge.

Murphy, Michelle. 2016a. "Alterlife in the Ongoing Aftermath: Exposure, Entanglement, Survivance." Toxic: A Symposium on Exposure, Entanglement, Survivance. Yale University, March 1. http://www.toxicsymposium.org/conversations-1/2016/3/1/alterlife-in-the-ongoing-aftermath-exposure-entanglement-survivance.

Murphy, Michelle. 2016b. "Expansive Affinities, Anti-Affinities, and Industrial Chemical Afterlife." In *4S/EASST 2016 Conference*. Barcelona. https://nomadit.co.uk/conference /easst2016/paper/30903.

Murphy, Michelle. 2017. "Alterlife and Decolonial Chemical Relations." *Cultural Anthropology* 32, no. 4 (November 20): 494–503. https://doi.org/10.14506/ca32.4.02.

Myers, Samuel S., Antonella Zanobetti, Itai Kloog, et al. 2014. "Increasing CO_2 Threatens Human Nutrition." *Nature* 510, no. 7503 (June 5): 139–42. https://doi.org/10.1038/ nature13179.

Nader, Laura. 2002. *The Life of the Law: Anthropological Projects*. Berkeley: University of California Press.

Nading, Alex M. 2017. "Local Biologies, Leaky Things, and the Chemical Infrastructure of Global Health." *Medical Anthropology* 36, no. 2 (February 17): 141–56. https://doi .org/10.1080/01459740.2016.1186672.

Nading, Alex M. 2020. "Living in a Toxic World." *Annual Review of Anthropology* 49, no. 49, (October 21): 209–24. https://doi.org/10.1146/annurev-anthro-010220 -074557.

Neuman, Johanna. 2006. "The Land of 10,770 Empty FEMA Trailers." *Los Angeles Times*, February 10. http://articles.latimes.com/2006/feb/10/nation/na-trailers10.

Nielsen, Marianne O. 1999. "Navajo Nation Courts, Peacemaking and Restorative Justice Issues." *Journal of Legal Pluralism and Unofficial Law* 31, no. 44 (January): 105–26. https://doi.org/10.1080/07329113.1999.10756539.

NOAA (National Oceanic and Atmospheric Administration). 2012. "2011 Tornado Information." *NOAA News*, March 20. http://www.noaanews.noaa.gov/2011_tornado _information.html.

Nold, Christian. 2017. "Device Studies of Participatory Sensing: Ontological Politics and Design Interventions." *UCL*. https://discovery.ucl.ac.uk/id/eprint/1569340/1 /Christian%20Nold%20Thesis.pdf.

Ntapanta, Samwel Moses. 2021. "'Lifescaping' Toxicants: Locating and Living with e-Waste in Tanzania." *Anthropology Today* 37, no. 4: 7–10. https://doi.org/10.1111/1467 -8322.12663.

NTP (National Toxicology Program). 2011. "Report on Carcinogens. Twelfth Edition, 2011." National Toxicology Program, Research Triangle Park, NC. https://ntrl.ntis.gov /NTRL/dashboard/searchResults/titleDetail/PB2011111646.xhtml.

Nunez, Yanelli, Jaime Benavides, Jenni A. Shearston, et al. 2024. "An Environmental Justice Analysis of Air Pollution Emissions in the United States from 1970 to 2010." *Nature Communications* 15, no. 1 (January 17): 268. https://doi.org/10.1038/s41467 -023-43492-9.

Odendahl, Marilyn. 2010. "Ohio-Based Greenlawn Homes Wins Bid for 15,000+ FEMA Units." *Elkhart Truth*, February 3. http://blogs.etruth.com/mainpoll/.

Offer, Avner. 1996. "The American Automobile Frenzy of the 1950s." In *Discussion Papers in Economic and Social History*. University of Oxford.

Ore, Janet. 2011. "Mobile Home Syndrome: Engineered Woods and the Making of a New Domestic Ecology in the Post–World War II Era." *Technology and Culture* 52, no. 2: 260–86. http://www.jstor.org/stable/23020568.

Oreskes, Naomi, and Erik M. Conway. 2011. *Merchants of Doubt: How a Handful of Scientists Obscured the Truth on Issues from Tobacco Smoke to Climate Change*. Reprint. New York: Bloomsbury.

Ottinger, Gwen. 2013. *Refining Expertise: How Responsible Engineers Subvert Environmental Justice Challenges*. New York: New York University Press.

Pacheco, Karin. 2009. "Clinical Summary: Cooper." May 18. National Jewish Hospital, Denver, CO.

Park, Kevin A. 2022. "Real and Personal: The Effect of Land in Manufactured Housing Loan Default Risk." *Cityscape* 24, no. 3: 339–62. https://www.jstor.org/stable/48707859.

Pasternak, Shiri. 2020. "Assimilation and Partition: How Settler Colonialism and Racial Capitalism Co-Produce the Borders of Indigenous Economies." *South Atlantic Quarterly* 119, no. 2 (April 1): 301–24. https://doi.org/10.1215/00382876-8177771.

Peña, Claudia. 2019. "Trauma Abounds: A Case for Trauma-Informed Lawyering." UCLA *Women's Law Journal* 26, no. 1: 7–16. doi:10.5070/l3261044345.

Pierce, Gregory, and Silvia Jimenez. 2015. "Unreliable Water Access in U.S. Mobile Homes: Evidence from the American Housing Survey." *Housing Policy Debate* 25, no. 4 (October 2): 739–53. https://doi.org/10.1080/10511482.2014.999815.

Pine, Jason. 2019. *The Alchemy of Meth: A Decomposition*. Minneapolis: University of Minnesota Press.

Povinelli, Elizabeth A. 2011. *Economies of Abandonment: Social Belonging and Endurance in Late Liberalism*. Durham, NC: Duke University Press.

Powell, Mark R. 1999. *Science at EPA: Information in the Regulatory Process*. Washington, DC: RFF Press.

Principles of Working Together Working Group. 1991. "People of Color Environmental Justice 'Principles of Working Together.'" Paper presented at the First People of Color Leadership Summit, Washington DC, October 27. https://www.ejnet.org/ej/workingtogether.pdf.

Proctor, Robert N., and Londa Schiebinger, eds. 2008. *Agnotology: The Making and Unmaking of Ignorance*. Stanford, CA: Stanford University Press.

Pulido, Laura. 2016. "Flint, Environmental Racism, and Racial Capitalism." *Capitalism Nature Socialism* 27, no. 3 (July 2): 1–16. https://doi.org/10.1080/10455752.2016.1213013.

Pulido, Laura. 2017. "Geographies of Race and Ethnicity II: Environmental Racism, Racial Capitalism and State-Sanctioned Violence." *Progress in Human Geography* 41, no. 4 (August): 524–33. https://doi.org/10.1177/0309132516646495.

Purifoy, Danielle M. 2020. "To Live and Thrive on New Earths." *Southern Cultures*. https://www.southerncultures.org/article/to-live-and-thrive-on-new-earths/.

Ray, Gene. 2004. "Reading the Lisbon Earthquake: Adorno, Lyotard, and the Contemporary Sublime." *Yale Journal of Criticism* 17, no. 1: 1–18. https://doi.org/10.1353/yale.2004.0007.

Red Nation. 2021. *The Red Deal: Indigenous Action to Save Our Earth*. Common Notions. https://www.commonnotions.org/the-red-deal.

Reno, Joshua. 2011. "Beyond Risk: Emplacement and the Production of Environmental Evidence." *American Ethnologist* 38, no. 3: 516–30. https://doi.org/10.1111/j.1548-1425.2011.01320.x

"Report of the Commissioner of Indian Affairs, 1877." 1877. H.R. Exec. Doc. 45th Cong., 2nd Sess. https://digitalcommons.law.ou.edu/cgi/viewcontent.cgi?article =6671&context=indianserialset.

Richter, Lauren, Alissa Cordner, and Phil Brown. 2018. "Non-Stick Science: Sixty Years of Research and (In)Action on Fluorinated Compounds." *Social Studies of Science* 48, no. 5 (October 1): 691–714. https://doi.org/10.1177/0306312718799960.

Roberts, Jody. 2010. "Reflections of an Unrepentant Plastiphobe: Plasticity and the STS Life." *Science as Culture* 19, no. 1 (March 1): 101–20. https://doi.org/10.1080 /09505430903557916.

Roberts, Jody. 2014. "Unruly Technologies and Fractured Oversight: Toward a Model for Chemical Control for the Twenty-First Century." In *Powerless Science? Science and Politics in a Toxic World*, edited by Soraya Boudia and Nathalie Jas. New York: Berghahn.

Robinson, Cedric J. 2000. *Black Marxism: The Making of the Black Radical Tradition*. 2nd ed. Chapel Hill: University of North Carolina Press.

Ross, Tracey, Chelsea Parsons, and Rebecca Vallas. 2016. "Creating Safe and Healthy Living Environments for Low-Income Families." Center for American Progress, July. https://www.americanprogress.org/wp-content/uploads/2016/07/SafeAndHealthy Homes-report.pdf.

Rozario, Kevin. 2007. *The Culture of Calamity: Disaster and the Making of Modern America*. Chicago: University of Chicago Press.

Rubaii, Kali. 2020. "Birth Defects and the Toxic Legacy of War in Iraq." MERIP, September 22. https://merip.org/2020/09/birth-defects-and-the-toxic-legacy-of -war-in-iraq/.

Russell, Dick, Sanford Lewis, and Brian Keating. 1992. "Inconclusive by Design: Waste, Fraud and Abuse in Federal Environmental Health Research." Environmental Health Network and National Toxics Campaign Fund, May.

Saadat, Sasan, Matt Vespa, and Mark Kresowik. 2020. "Rhetoric vs. Reality: The Myth of 'Renewable Natural Gas' for Building Decarbonization." *EarthJustice*, July. https:// earthjustice.org/sites/default/files/feature/2020/report-decarb/Report_Building -Decarbonization-2020.pdf.

Safecast. n.d. "About—Safecast." Accessed October 31, 2022. https://safecast.org /about/.

Salthammer, Tunga, Sibel Mentese, and Rainer Marutzky. 2010. "Formaldehyde in the Indoor Environment." *Chemical Reviews* 110, no. 4 (April 14): 2536–72. https://doi .org/10.1021/cr800399g.

Sapien, Joaquin. 2010. "Senior Public Health Official Reassigned in Wake of Congressio- nal Inquiries." *ProPublica*, January 22. http://www.propublica.org/feature/senior-cdc -official-reassigned-howard-frumkin.

Saxton, Dvera I. 2021. *The Devil's Fruit: Farmworkers, Health, and Environmental Justice*. New Brunswick, NJ: Rutgers University Press.

Schrader, Astrid. 2010. "Responding to Pfiesteria Piscicida (the Fish Killer) Phantomatic Ontologies, Indeterminacy, and Responsibility in Toxic Microbiology." *Social Studies of Science* 40, no. 2: 275–306. https://doi.org/10.1177/0306312709344902.

Schroeder, Tim, David Bond, and Janet Foley. 2021. "PFAS Soil and Groundwater Contamination via Industrial Airborne Emission and Land Deposition in SW Vermont and Eastern New York State, USA." *Environmental Science: Processes and Impacts* 23, no. 2 (March 4): 291–301. https://doi.org/10.1039/D0EM00427H.

Seed Commons. n.d. "About Seed Commons." Accessed August 31, 2022. https://seedcommons.org/about-seed-commons/.

Seltenrich, Nate. 2012. "Healthier Tribal Housing: Combining the Best of Old and New." *Environmental Health Perspectives* 120, no. 12 (December): a460–69. https://doi.org/10.1289/ehp.120-a460.

Shadaan, Reena. 2023. "Healthier Nail Salons: From Feminized to Collective Responsibilities of Care." *Environmental Justice* 16, no. 1 (February): 62–71. https://doi.org/10.1089/env.2021.0097.

Shadaan, Reena, and Michelle Murphy. 2020. "EDCs as Industrial Chemicals and Settler Colonial Structures: Towards a Decolonial Feminist Approach." *Catalyst* 6, no. 1 (May 15). https://doi.org/10.28968/cftt.v6i1.32089.

Shapiro, Nicholas. 2013. "Nick Shapiro: Dreamscapes of Dispossession." *Guernica*, February 26. https://www.guernicamag.com/nick-shapiro-landscapes-of-dispossession/.

Shapiro, Nicholas. 2014a. "DIY Formaldehyde Test Kit." Public Lab, November 3. https://publiclab.org/n/11317.

Shapiro, Nicholas. 2014b. "DIY Indoor Air Quality Remediation." Public Lab, October 20. https://publiclab.org/n/11282.

Shapiro, Nicholas. 2015. "Attuning to the Chemosphere: Domestic Formaldehyde, Bodily Reasoning, and the Chemical Sublime." *Cultural Anthropology* 30, no. 3 (August 10): 368–93. https://doi.org/10.14506/ca30.3.02.

Shapiro, Nicholas. 2019a. "Manufacturing Home." *Journal for the Anthropology of North America* 22, no. 2: 121–24. https://doi.org/10.1002/nad.12113.

Shapiro, Nicholas. 2019b. "Persistent Ephemeral Pollutants." In *Being Material*, edited by Marie-Pier Boucher, Stefan Helmreich, Rebbeca Uchill, and Leila Kinney, 154–61. Cambridge, MA: MIT Press.

Shapiro, Nicholas, and Gretchen Gehrke. 2015. "Community Formaldehyde Monitoring at Natural Gas Compressor Stations, Protocol and Data Sheet." Public Lab, May 8. https://publiclab.org/notes/nshapiro/05-08-2015/community-formaldehyde-monitoring-at-natural-gas-compressor-stations-protocol-and-data-sheet.

Shapiro, Nicholas, Gretchen Gehrke, Liz Barry, and Silent Disaster. 2016. "Plant Air Remediation Field Test #2 (Georgia)." Public Lab. https://publiclab.org/n/13325.

Shapiro, Nicholas, Nasser Zakariya, and Jody Roberts. 2017. "A Wary Alliance: From Enumerating Environments to Inviting Apprehension." *Engaging STS* 3: 575–602. https://doi.org/10.17351/ests2017.133.

Shehab, Nadine, Michael P. Anastario, and Lynn Lawry. 2008. "Access to Care Among Displaced Mississippi Residents in FEMA Travel Trailer Parks Two Years After Katrina." *Health Affairs* 27, no. 5 (September): w416–29. https://doi.org/10.1377/hlthaff.27.5.w416.

Simmons, Kristen. 2017. "Settler Atmospherics." Member Voices, *Society for Cultural Anthropology*, November 20. https://culanth.org/fieldsights/1221-settler-atmospherics.

Simpson, Leanne Betasamosake. 2017. *As We Have Always Done: Indigenous Freedom Through Radical Resistance*. 3rd ed. Minneapolis: University of Minnesota Press.

Singh, Jewellord Nem. 2021. "Mining Our Way Out of the Climate Change Conundrum? The Power of a Social Justice Perspective." Wilson Center, October. https://www.wilsoncenter.org/sites/default/files/media/uploads/documents/Mining%20Our%20Way%20Out%20of%20the%20Climate%20Change%20Conundrum%20The%20Power%20of%20a%20Social%20Justice%20Perspective%20by%20Jewellord%20Nem%20Singh%2011.1.pdf.

Sismondo, Sergio. 2018. *Ghost-Managed Medicine: Big Pharma's Invisible Hands*. Mattering Press.

60 Minutes. 2019. "Lumber Liquidators to Pay $33 Million Criminal Penalty," March 1. https://www.cbsnews.com/news/lumber-liquidators-to-pay-33-million-criminal-penalty-60-minutes/.

Sobrevila, Claudia. 2008. *The Role of Indigenous Peoples in Biodiversity Conservation: The Natural but Often Forgotten Partners*. World Bank. https://web.archive.org/web/20120802223531/https://siteresources.worldbank.org/intbiodiversity/Resources/RoleofIndigenousPeoplesinBiodiversityConservation.pdf.

Solnit, Rebecca. 2009. *A Paradise Built in Hell: The Extraordinary Communities That Arise in Disasters*. New York: Penguin.

Soret, A., M. Guevara, and J. M. Baldasano. 2014. "The Potential Impacts of Electric Vehicles on Air Quality in the Urban Areas of Barcelona and Madrid (Spain)." *Atmospheric Environment* 99 (December 1): 51–63. https://doi.org/10.1016/j.atmosenv.2014.09.048.

Spackman, Christy. 2020. "In Smell's Shadow: Materials and Politics at the Edge of Perception." *Social Studies of Science* 50, no. 3 (June 1): 418–39. https://doi.org/10.1177/0306312720918946.

Spade, Dean. 2016. *Coming Together Speaker Series: Dean Spade*. YouTube video, 45:21, May 4. https://www.youtube.com/watch?v=D1HtLMi-ELU.

Spade, Dean. 2020. *Mutual Aid: Building Solidarity During This Crisis (and the Next)*. London: Verso.

Spiegel, Alix. 2007. "Stuck and Suicidal in a Post-Katrina Trailer Park." *NPR*, August 8. http://www.npr.org/templates/story/story.php?storyId=12592168.

Stewart, Kathleen. 2005. "Cultural Poesis: The Generativity of Emergent Things." In *Handbook of Qualitative Research*, edited by Norman Denzin and Yvona Lincoln, 3rd ed., 1027–42. London: Sage.

Stewart, Kathleen. 2007. *Ordinary Affects*. Durham, NC: Duke University Press.

Strum, Madeleine, and Richard Scheffe. 2016. "National Review of Ambient Air Toxics Observations." *Journal of the Air and Waste Management Association* 66, no. 2 (February 1): 120–33. https://doi.org/10.1080/10962247.2015.1076538.

Stutte, Gary W. 2012. "Phytoremediation of Indoor Air: NASA, Bill Wolverton, and the Development of an Industry." January 1. http://ntrs.nasa.gov/search.jsp?R=20120003454.

Subcommittee on Investigations and Oversight. 2009. "The Agency for Toxic Substances and Disease Registry (ATSDR): Problems in the Past, Potential for the Future?" Committee on Science and Technology, US House of Representatives, March 10.

Sullivan, Esther. 2014. "Halfway Homeowners: Eviction and Forced Relocation in a Florida Manufactured Home Park." *Law and Social Inquiry* 39, no. 2 (April): 474–97. https://doi.org/10.1111/lsi.12070.

Sullivan, Esther. 2018. *Manufactured Insecurity: Mobile Home Parks and Americans' Tenuous Right to Place*. Berkeley: University of California Press.

Sullivan, Esther, Carrie Makarewicz, and Andrew Rumbach. 2021. "Affordable but Marginalized." *Journal of the American Planning Association* (August 25): 1–13. https://doi.org/10.1080/01944363.2021.1952477.

Tabuchi, Hiroko, and Danny Hakim. 2016. "How the Chemical Industry Joined the Fight Against Climate Change." *New York Times*, October 17. https://www.nytimes.com/2016/10/17/business/how-the-chemical-industry-joined-the-fight-against-climate-change.html.

Táíwò, Olúfẹ́mi O. 2022. *Elite Capture: How the Powerful Took Over Identity Politics (and Everything Else)*. Chicago: Haymarket Books.

TallBear, Kim. 2014. "Standing with and Speaking as Faith: A Feminist-Indigenous Approach to Inquiry." *Journal of Research Practice* 10, no. 2 (July): 1–7. https://search.ebscohost.com/login.aspx?direct=true&db=a9h&AN=101850329&site=ehost-live.

Tan, Tao, Yun Zhang, Wenhong Luo, et al. 2018. "Formaldehyde Induces Diabetes-Associated Cognitive Impairments." *FASEB Journal* 32, no. 7: 3669–79. https://doi.org/10.1096/fj.201701239R.

Taylor, Keeanga-Yamahtta. 2019. *Race for Profit: How Banks and the Real Estate Industry Undermined Black Homeownership*. Illustrated ed. Chapel Hill: University of North Carolina Press.

TenHoor, Meredith, and Jessica Varner. 2023. "Mattering Toxics and Making Toxics Matter in Architecture and Landscape Histories." *Aggregate* 11 (May). https://doi.org/10.53965/WADN3098.

Thylstrup, Nanna Bonde. 2019. "Data Out of Place: Toxic Traces and the Politics of Recycling." *Big Data and Society* 6, no. 2 (July 1). https://doi.org/10.1177/2053951719875479.

Tironi, Manuel. 2018. "Hypo-Interventions: Intimate Activism in Toxic Environments." *Social Studies of Science* 48, no. 3 (June 1): 438–55. https://doi.org/10.1177/0306312718784779.

Tonn, B., E. Rose, and B. Hawkins. 2018. "Evaluation of the U.S. Department of Energy's Weatherization Assistance Program: Impact Results." *Energy Policy* 118 (July 1): 279–90. https://doi.org/10.1016/j.enpol.2018.03.051.

Truong, Thanh. 2013. "FEMA Sends Extra Settlement Checks to Plaintiffs in Formaldehyde Suit." *WWLTV*, November 6. http://www.wwltv.com/news/FEMA-sends-extra-settlement-checks-to-plaintiffs-in-formaldehyde-suit-230909371.html.

Tso, Tom. 1992. "Moral Principles, Traditions, and Fairness in the Navajo Nation Code of Judicial Conduct." *Judicature* 76, no. 1: 15–21. https://heinonline.org/HOL/P?h=hein.journals/judica76&i=17.

Tuck, Eve. 2009. "Suspending Damage: A Letter to Communities." *Harvard Educational Review* 79, no. 3 (September 1): 409–28. https://doi.org/10.17763/haer.79.3.n0016675661t3n15.

Tuck, Eve, and K. Wayne Yang. 2016. "What Justice Wants." *Critical Ethnic Studies* 2, no. 2: 1–15. https://doi.org/10.5749/jcritethnstud.2.2.0001.

Tulpule, Ketki, and Ralf Dringen. 2013. "Formaldehyde in Brain: An Overlooked Player in Neurodegeneration?" *Journal of Neurochemistry* 127, no. 1: 7–21. https://doi.org/10.1111/jnc.12356.

Turner, Margery Austin. 2007. "Alternative to the FEMA Trailer Parks: Lessons from Social Science Research." Statement before the Committee on Transportation and Infrastructure, Subcommittee on Economic Development, Public Buildings and Emergency Management, US House of Representatives, Washington, DC, March 20.

Updike, John. 1960. *Rabbit, Run*. Reissue. New York: Random House.

US Criminal Code. n.d. Claims Under the Federal Tort Claims Act, 32 CFR Part 842 Subpart I US §. Accessed September 5, 2022. https://www.ecfr.gov/current/title-32/subtitle-A/chapter-VII/subchapter-D/part-842/subpart-I.

US Department of Housing and Urban Development. 2024. "American Healthy Homes Survey II Additional Environmental Findings." Washington, DC.

Van Herk, Aritha. 1991. "Post-Modernism: Homesick for Homesickness." In *The Commonwealth Novel Since 1960*, edited by Bruce King, 216–30. London: Palgrave Macmillan.

Varner, Jessica. Forthcoming. "Chemical Desires: When the Chemical Industry Met Modern Design (1870–1970)." Chicago: University of Chicago Press.

Vera, Lourdes. 2022. "Environmental Data Justice in Action: Civically Valid Air Monitoring Near Oil and Gas Extraction in the Eagle Ford Shale Play." PhD diss., Northeastern University. https://search.proquest.com/openview/76741bb0e378a30722eb96c3162e9bfc/1?pq-origsite=gscholar&cbl=18750&diss=y.

Vera, Lourdes, Garance Malivel, Drew Michanowicz, Choong-Min Kang, and Sara Wylie. 2020. "Photopaper as a Tool for Community-Level Monitoring of Industrially Produced Hydrogen Sulfide and Corrosion." *Atmospheric Environment* X5 (January 1). https://doi.org/10.1016/j.aeaoa.2019.100049.

Verdin, Monique Michelle. 2019. *Return to the Yakni Chitto: Houma Migrations*. Edited by Rachel Breunlin. New Orleans, LA: The Neighborhood Story Project, an imprint of University of New Orleans Press.

Voyles, Traci Brynne. 2015. *Wastelanding: Legacies of Uranium Mining in Navajo Country*. Minneapolis: University of Minnesota Press.

Wakefield-Rann, Rachael. 2021. *Life Indoors: How Our Homes Are Shaping Our Bodies and Our Planet*. London: Palgrave Macmillan.

Wang, Hai-xu, He-cheng Li, Mo-qi Lv, et al. 2015. "Associations Between Occupation Exposure to Formaldehyde and Semen Quality, a Primary Study." *Scientific Reports* 5, no. 1 (October 30). https://doi.org/10.1038/srep15874.

Wang, Z., J. C. DeWitt, C. P. Higgins, and I. T. Cousins. 2017. "A Never-Ending Story of Per- and Polyfluoroalkyl Substances (PFASS)?" *Environmental Science and Technology* 51, no. 5 (March): 2508–18. https://doi.org/10.1021/acs.est.6b04806.

Wantke, F., C. M. Demmer, P. Tappler, M. Götz, and R. Jarish. 1996. "Exposure to Gaseous Formaldehyde Induces IgE-Mediated Sensitization to Formaldehyde in School-

Children." *Clinical and Experimental Allergy* 26, no. 3: 276–80. https://doi.org/10.1111
/j.1365-2222.1996.tb00092.x.

Weizman, Eyal. 2011. *The Least of All Possible Evils: Humanitarian Violence from Arendt to Gaza.* New York: Verso.

Wells, Katie J., Kafui Attoh, and Declan Cullen. 2021. "'Just-in-Place' Labor: Driver Organizing in the Uber Workplace." *Environment and Planning A: Economy and Space* 53, no. 2 (March 1): 315–31. https://doi.org/10.1177/0308518X20949266.

Western, Bruce, and Katherine Beckett. 1999. "How Unregulated Is the U.S. Labor Market? The Penal System as a Labor Market Institution." *American Journal of Sociology* 104, no. 4: 1030–60. https://doi.org/10.1086/210135.

White, Monica M. 2018. *Freedom Farmers: Agricultural Resistance and the Black Freedom Movement.* Chapel Hill: University of North Carolina Press.

Whitmarsh, Ian. 2008. *Biomedical Ambiguity: Race, Asthma, and the Contested Meaning of Genetic Research in the Caribbean.* Ithaca, NY: Cornell University Press.

Whyte, Kyle. 2012. "Indigenous Peoples, Solar Radiation Management, and Consent." In *Engineering the Climate: The Ethics of Solar Radiation Management*, edited by Christopher J. Preston, 65–76. Lanham, MD: Lexington Books.

Wilkinson, Annabelle. 2020. "Racial Disparity in Manufactured Housing: A Study of Affordability in the United States." Michigan State University. https://d.lib.msu.edu /etd/48471/datastream/OBJ/view.

Winkler-Schmit, David. 2006. "Marlyville / Fontainebleau / Broadmoor After Hurricane Katrina—News Clippings Page." *New Orleans Gambit*, March 14. http://www .vendomeplace.org/press031806gambitbroadmoor.html.

Wolfson, Mariel. 2013. "Defining Healthy Housing: The Competing Priorities of Energy and Indoor Air Quality in the Early 1980s." June. Joint Center for Housing Studies, Harvard University. https://www.jchs.harvard.edu/sites/default/files/media/imp/w13 -4_wolfson.pdf.

Wolverton, B. C. 2010. "Plant-Based Air Filters for Formaldehyde Remediation in FEMA Trailers." Picayune, MS: Wolverton Environmental Services.

Wool, Zoë. 2011. "Miniscule War: The Common Sense of Fragments at Walter Reed Army Medical Center." Presentation at the American Anthropological Association, San Francisco, November 19.

Wylie, Sara Ann. 2018. *Fractivism: Corporate Bodies and Chemical Bonds.* Durham, NC: Duke University Press.

Wylie, Sara Ann, Nicholas Shapiro, and Max Liboiron. 2017. "Making and Doing Politics Through Grassroots Scientific Research on the Energy and Petrochemical Industries." *Engaging STS* 3: 393–425. https://doi.org/10.17351/ests2017.134.

Zachria, Anthony, and Bela Patel. 2006. "Deaths Related to Hurricane Rita and Mass Evacuation." *CHEST Journal* 130 (October 1): 124S.

Zee, Jerry C. 2017. "Holding Patterns: Sand and Political Time at China's Desert Shores." *Cultural Anthropology* 32, no. 2 (May 1): 215–41. https://doi.org/10.14506 /ca32.2.06.

Zee, Jerry C. 2022. *Continent in Dust: Experiments in a Chinese Weather System.* Berkeley: University of California Press.

Zeng, Yicheng, Prashik Manwatkar, Aurélie Laguerre, et al. 2021. "Evaluating a Commercially Available In-Duct Bipolar Ionization Device for Pollutant Removal and Potential Byproduct Formation." *Building and Environment* 195 (May 15). https://doi.org/10.1016/j.buildenv.2021.107750.

Zhang, Siyang, Nicholas Shapiro, Gretchen Gehrke, et al. 2019. "Smartphone App for Residential Testing of Formaldehyde (SmART-Form)." *Building and Environment* 148 (January 15): 567–78. https://doi.org/10.1016/j.buildenv.2018.11.029.

Zhu, Lei, D. Jacob, L. Mickley, et al. 2017. "Observing Formaldehyde (HCHO) from Space: Trend Analysis and Public Health Implications." Poster, Cambridge, MA, May 1.

Zion, James, and Robert Yazzie. 1997. "Indigenous Law in North America in the Wake of Conquest." *Boston College International and Comparative Law Review* 20, no. 1 (December 1): 55. https://heinonline.org/HOL/P?h=hein.journals/bcic20&i=63.

INDEX

Italic page numbers refer to figures

Colorado, 95
commodity fetishism, 63–64
community science, 82, 140–41
community wealth cooperatives, 18, 167
congestive heart failure, 107–8
congressional hearings, xi, 4, 88
Cooper, Christopher, 89–91, 95–97
cooperatives, 18, 20, 166–67
COP21 Solutions conference, 164–65
Costelloe-Koehn, Brandon, 22
Craigslist, 50
Crayola, 61
Cree, Delvin, 36–37
Critical Resistance (CR), 160

Dalton, Pamela, 188n9
Daniella, 106, 109–10, 115
Daniels, Francesca, 109, 190n14
Dannemiller, Karen, 137
Daryl, 42–43
data treadmills, 10, 160
Dave, 107–8
Democratic Republic of the Congo, 196n17
Denorex, 57
Department of Housing and Urban Development (HUD), 16, 100, 144, 154, 183n4, 195n10
depression, xiii, xviii–xix, 89
detoxification, 9, 12, 111, 136, 152, 171. *See also* remediation
diabetes, 36
Diplomate of the American Board of Toxicology (DABT), 109
disability, xix, 24, 114, 187n2. *See also* chronic illness
disaster capitalism, 13, 80
Disaster Relief Act, 183n4
displacement, xv–xvii, 12, 90, 117, 123, 183n4, 184n5
distributed harm, 22, 65–66, 69, 77
documentation, xi, xviii, 4–5, 82, 89, 127, 136, 139, 142, 146, 149, 186n13; of harm, 6–10, 95, 108–9, 127, 127–32, 129, 151, 160
dogs, 38, 71, 73–75, 136, 188n12
dog whistle racism, 43
domestic violence, xvi
Donahue, Allie, 137
Dow Chemical, 59

Dulce Lomita Mobile Home Cooperative, 167
DuPont, 80

East Jefferson General Hospital, 91
eBay, 44–45
eczema, xv, 56
Elver, Rikki, 115–17
energy efficiency, 14, 70, 188n16
engineered wood, 1, 16, 19–20, 24, 44, 50, 68, 71, 80, 141, 155, 198n29. *See also* manufactured homes
England: London, 8, *63*
Enlightenment, 66
enumeration, 12, 131, 145, 147–48, 152
Environmental Health Network, 87
environmental injustice, 5, 7–13, 155–60
environmentalism, 6–7, 9, 156, 159–61, 167
environmental justice, 6–7, 13, 22, 110, 130, 146, 161, 192n12, 196n12
Environmental Protection Agency (EPA), xviii, 150, 153, 193n3, 194nn1–2, 195n4, 195n9; Clean Air Act Advisory Committee, 195n8; exposure thresholds, 2, 81, 84, 106, 180; hazard testing, 83–88; Indoor Air Division, 131; Integrated Risk Information System (IRIS), 149, 155, 179; Office of Toxic Substances, 154
ethics, 6, 57, 94, 115, 127, 130, 135, 145, 148, 194n14
ethnography, 6, 21, 33, 119, 140
Europe, 13, 15, 24, 31, 58, 62, 63, 69, 156
European Union, 156
eventfulness, 9–10, 76, 103, 127, 185n7
evidence, 5, 9–12, 89, 94, 115, 127, 146, 186n17, 193n1; bodies as, 71–76
exacerbation, 3, 20, 89, 91–93, 95–96, 192n5
expert witnesses, 89
exposure, 6–7, 20, 23–24, 30–32, 47–48, 50, 65, 122, 149, 151, 153, 155–56, 159, 167, 171, 185n7, 189n4, 190n14, 193n2; and bodily attunement, 66–69, 71–76; chronic, xi, xvii, 4, 10, 38, 56, 69–70, 81, 102–3, 109, 189n9, 195n10; disavowal of, 40; inequities of, 170; and late industrial sublime, 69–71; and litigation, 12, 102–5, 108–11, 116, 118; long-term effects of, 51–53; mitigation of, 128–48, 191n1; in mobile homes, 18;

laminated wood, 2, 16, 80, 143, 156, 196n14

LA Moves, 184n6

#LandBack, 168

Land Institute, 168

Lara, Elizabeth, 137

late industrial sublime, 57, 65–66, 68–71, 74, 76–77, 97

Latinx people, 15, 17–18, 44, 155

Lawrence Berkeley National Laboratory, 50

Lea, Tess, 33

legislation, 4, 12, 150–51, 153, 157. *See also individual laws*

Lerner, Sharon, 195n4

lesser evil, 151–52, 159

level of concern, 84, 87–88, 189n11

Lezaun, Javier, 190n16

liability, 12, 28, 37, 44, 83, 118, 120–21; federal, 35–36, 40, 85, 88, 117; mitigation of, 32; product liability court cases, 21, 23, 103, 116; seller, 35. *See also* accountability; responsibility

Liboiron, Max, 24, 53, 161, 163

Lippencott, Mathew, 142

litigation, 12, 23, 81–83, *132*, 152, 154, 169, 185n6, 192n10; accountability in, 99–124, 129, 185n8; asthma in, 88–97, 190n14; bellwether trials, 89, 117; class-action litigation, 89, 116–17, 196n14; frivolous lawsuits, 118; risk of, 45, 71; settlements, 103, 111, 116, 120–22, 129, 191, 192nn6–7, 192nn11–12; and theorizing change, 125, 127–29; torts, 105, 119–24, 129, 192n9; toxic tort litigation, 5, 111, 114–18. *See also* liability

Lizzie, xvi, 55–57, 112–15, 119–20, 122–23, 169, 192n11

loans, 14, 16–17, 167, 186n15. *See also* mortgages

Louisiana, 23, 110, 117, 125, 168, 183n3; Baton Rouge, xiii; Boutte, xiii; and dog whistle racism, 43; floodplains, 33, 38; Hurricane Katrina damage in, xi–xvii; Jefferson, 81; Livingston, 45; New Sarpy, 147. *See also* New Orleans, LA

Louisiana Department of Economic Development, 110; Early Return Program, xii

"Louisiana people" dog whistle, 43

Louisiana Supreme Court, 110

Luan, 1, 80

Lumber Liquidators (LL), 143, 156, 196n14; stock shorting, 156

Luxemburg, Rosa, 185n5

Mac, 106–11, 114, 120, 122–24

making and doing, 24

manufactured homes, 13, 15–18, 31–32, 35, 43, 45, 62, 143, 167, 186nn14–15, 195n10. *See also* engineered wood

manufacturers, 17, 62, 64, 80–81, 89, 108, 115–20, 155, 190n13, 192n6, 195n4

Marcellus shale field, 8

Marquisha, xiv–xv

Marx, Karl, 63

Masco, Joseph, 10, 69

Massachusetts: Cambridge, 149

Maygarden, Benny, ix, 183n2

McDonald's coffee lawsuit, 118, 192n8

McFeely, Dick, 71, 188n18

McFeely, Harriett, 71–75, 77, 129, 185n6, 188nn19–20

McGoey, Linsey, 189n4

Meikle, Jeffrey, 64

Melody, 79

Meredith, 44–45, 187n3

Merry, Sally Engle, 119

methane, 3, 158, 195n6

methodology of book, 3, 5–7, 21–24, 186n13. *See also* ethnography

Metzl, Jonathan, 14

Michigan, 56, 187n1; Flint, 122, 185n10, 192n12

microemissions, 2, 76

Middle East, 24

Miller, Brad, 87

Miranda, xvi

misogynoir, 96–97

Mississippi, 27, 30, 106, 117, 125, 131, 135–36, 193n1; Bay St. Louis, 79; Biloxi, 81; FEMA mobile homes in, 38–*39*, 55, 79–81, 112–*13*; FEMA trailers in, xvii–xviii, 21, 23, 33; Hurricane Katrina damage in, xii, xv–xviii, 55, 79; Jackson, 120; Pass Christian, 55, 112; Picayune, 38–*39*, 127, 136; Purvis, 36

Mississippi River, xiii

Missouri, 44, 150, 187n3; St. Louis, 20

MIT, 139, 164

mitigation, 8, 12, 20, 32, 50, 71, 77, 130–31, 140, 159, 167, 184n6, 185n5, 191n1

US National Weather Service: Global
Forecast System, 197n27
US Supreme Court, 150, 193n1

Vanderbilt Mortgage, 17
ventilation, 2, 38, 44k ventilation83, 86, 88,
136, 151, 155, 190n13
Vera, Lourdes, 146
Verdin, Monique: Land Memory Bank and
Seed Exchange, 168
Veterans Affairs, 49
Vietnam, xvi
Vietnam Church, xvi
Vietnamese Americans, xvii
Vietnamese people, xvi
Virginia, 27
volatile organic chemicals (VOCs), 56, 62, 126,
134–35, *142*, 148, 187n3, 194nn9–11

Walmart, 46, 61
warnings, 12, 28–30, 32, *34*–35, 41, 187n2
Washington State, 36
Weizman, Eyal, 159
welfare queen stereotype, 43
West Virginia, 46, 56, 168
wheel estate: *vs.* real estate, 14, 20
whiteness, xii, 119, 168, 189n3; and dog whistle
racism, 43–44; exclusion from, 64; and

housing, 12–20; and Hurricane Katrina
damage, xv, 12; and litigation, 91, 123;
and methodology of book, 21; and racial
segregation, xvii; relation to Blackness,
185n10; and trailer recipients, xv–xvi, xviii,
12–15, 31
white supremacy, 31, 123, 192n12, 193n1.
See also Jim Crow; racism; Southern
Manifesto
"white trash" stereotype, 13
white underclass, 12
Whitmarsh, Ian, 95
Williams, Patricia, 93–96
WLOX, 81
Wolverton, Bill, 126, 134–35, 137
Wool, Zoë, 183n2
Woolgar, Steve, 190n16
World Health Organization (WHO), 41, 75,
180, 195n10
World War I, 31
World War II, 58, 64, 95; Japanese American
internment, 184n7
Wright, Richard, 31
Wylie, Sara, 194n7

Yang, K. Wayne, 122

Zakariya, Nasser, 197n23

www.ingramcontent.com/pod-product-compliance
Lightning Source LLC
Chambersburg PA
CBHW020852270326
41928CB00006B/667